Ultrafine Bubbles

Ultrafine Bubbles

edited by

Koichi Terasaka | Kyuichi Yasui
Wataru Kanematsu | Nobuhiro Aya

JENNY STANFORD
PUBLISHING

Published by

Jenny Stanford Publishing Pte. Ltd.
Level 34, Centennial Tower
3 Temasek Avenue
Singapore 039190

Email: editorial@jennystanford.com
Web: www.jennystanford.com

British Library Cataloguing-in-Publication Data
A catalogue record for this book is available from the British Library.

Ultrafine Bubbles

ISBN 978-981-4877-59-6 (Hardcover)
ISBN 978-1-003-14195-2 (eBook)

Contents

Preface

Ultrafine bubbles (UFBs) are defined as gas (and vapor) bubbles smaller than 1 μm in diameter. Sometimes they are called bulk nanobubbles. However, the upper limit of the size of UFB (a bulk nanobubble) is one order of magnitude larger than that of a nanoparticle, which is 100 nm in diameter. In other words, the prefix "nano-" for bubbles means one order of magnitude larger than "nano-" for particles. It may be confusing. Thus, we call those bubbles UFBs instead of bulk nanobubbles. Furthermore, in ISO (International Organization for Standardization)/TC 281 (technical committee: Fine Bubble Technology), bubbles smaller than 1 μm in diameter are defined as UFBs. Bubbles larger than 1 μm and smaller than 100 μm are defined as microbubbles. UFBs and microbubbles are called fine bubbles.

Although UFBs have already been used in commercial processes such as cleaning, plant cultivation, and others, there are many mysteries in UFBs such as mechanisms of stability, radical formation, biological and medical effects, etc. In this book, reviews on UFB research are provided for readers who are interested in UFB fundamentals and their applications, including cleaning, biological, medical, and dental ones.

Chapters 2 and 7 are basically translations of the chapters in the Japanese book titled *Introduction to Fine Bubble Science and Technology* edited by the Union of Fine Bubble Scientists and Engineers and published by Nikkan Kogyo Shimbun, Ltd., Tokyo, in 2016.

We hope that this book is useful for many students and researchers who are interested in this newly developing field.

Koichi Terasaka
Kyuichi Yasui
Wataru Kanematsu
Nobuhiro Aya
April 2021

Chapter 1

History of Ultrafine Bubbles

Akira Yabe
National Institute of Advanced Industrial Science and Technology (AIST), Tsukuba, Ibaraki 305-8569, Japan
Technology Strategy Center, New Energy and Industrial Technology Development Organization (NEDO), Kawasaki, Kanagawa 212-8554, Japan
yabe-akira@aist.go.jp; yabeakr@nedo.go.jp

In this chapter, the history of ultrafine bubbles is discussed from various aspects. Starting from the actual applications of microbubbles in the 20th century for froth flotation, purification of contaminated water in lakes, and the enhancement of growth of living things in oceans, the various kinds of effects of microbubbles and ultrafine bubbles have been explained historically. Since the ultrapure water has become one of the key technologies for the semiconductor process, the contaminated solid particles in the ultrapure water have become negligibly small amount. Therefore, the ultrafine bubbles having diameter below 1 μm have become the targets of researches without a large number of contaminated solid particles. Many challenges have tried to clear the characteristics of ultrafine bubbles and to create the useful effects of ultrafine bubbles. Furthermore,

Ultrafine Bubbles
Edited by Koichi Terasaka, Kyuichi Yasui, Wataru Kanematsu, and Nobuhiro Aya

ISBN 978-981-4877-59-6 (Hardcover), 978-1-003-14195-2 (eBook)
www.jennystanford.com

the experimental evidence of ultrafine bubbles and the theoretical investigation of the mechanism of existing ultrafine bubbles for the long time have been tried actively so far and explained historically.

1.1 Characteristics of Ultrafine Bubbles, Microbubbles, and Fine Bubbles

Ultrafine bubbles are floating gas bubbles in liquids and have diameters below 1 μm. Microbubbles have diameter between 1 μm and 100 μm. Ultrafine bubbles and microbubbles are called part of fine bubbles. These terminologies have been determined by the ISO (International Standard Organization) TC281 Fine Bubble technology as ISO 20480-1 in 2017. Before this determination, the ultrafine bubbles have been frequently called "nano bubbles" since these have the diameter of hundreds of nanometers. However, in ISO TC229 TS 80004-1, "nano objects" have been specified as having the length scale below 100 nm; the expression of "nano bubbles" became inadequate for nanoobjects from the specification of ISO TS 80004-1. Therefore, the expression of ultrafine bubbles has been specified in ISO TC-281, and ultrafine bubbles have been utilized widely around the world.

1.2 History of Microbubble Applications

Because of their visibility, microbubbles have been actually introduced in various application areas since the 20th century. The typical application areas are froth flotation, ultrasonic imaging, purification of contaminated water, and the enhancement of growth of living beings in the ocean.

1.2.1 Froth Flotation

As a commercial and useful application of microbubbles, the froth flotation method of producing various kinds of metals such as zinc was established in 1905 at Broken Hill in Australia. This floating method utilized microbubbles and surfactants based on the mechanism that the bubble surfaces would attract the hydrophobic metallic element more than the hydrophilic rock element. The mean bubble size would range between 50 μm and 650 μm [1].

1.2.2 Fine Bubbles for Ultrasonic Imaging inside Body

Ultrasonic contrast agents for medical application have been established from the viewpoints of fine bubble formation and microcapsule generation [2, 3]. The diameter of fine bubbles is about 2 μm and the surfactant is frequently absorbed on the surface.

1.2.3 Purification of Contaminated Water

Cleaning and purification of contaminated water in lakes, ponds, and dams with oil contamination and lack of oxygen have been carried out since about 1995 [4]. The swirling flow bubble generators created microbubbles having diameters of about 20 μm.

1.2.4 Enhancement of Growth of Oysters, Scallops, and Pearls

These technologies were established for getting rid of the harmful algal blooms in 1999 [5]. These algal blooms had made the color of sea water red and decreased the level of oxygen for fishes and oysters. The injection of microbubbles-included water decreased the significant effects for damages. Various application areas have been developed such as effective bathing and drag reduction of ships by utilizing microbubbles having diameter between 10 μm and 30 μm.

1.3 Historical Background of Ultrafine Bubbles

As the historical background of fine bubbles and ultrafine bubbles, the great progress in the academic research of two-phase flow and the tremendous progress in the ultrapure water production system have been very important.

1.3.1 History of Academic Researches of Fine Bubbles in Two-Phase Flow

The academic progress of the two-phase flow research has been divided into four stages [6]: 1948–59, 1960–70, 1971–79, and 1980–88. The bubble flows have been researched from the first stage. The flow patterns of two-phase flow have been categorized, and their transitions have been discussed for bubble flow by the void fraction behaviors. The phenomena relating to fine bubbles and ultrafine bubbles would be in the category of bubble flow. The characteristics of bubbles have been researched in the latter half of the 20th century, and the systematic understanding of bubbles has already been established [7–9]. However, at that time, bubbles having diameter between several mm and 10 cm were utilized largely in chemical engineering applications. Smaller-diameter bubbles such as microbubbles have not yet been researched so far, since the generation of smaller bubbles is necessary to use a large amount of energy [10]. Furthermore, since microbubbles and ultrafine bubbles are rather new phenomena that appeared around 1985 and are complicated, including the effects of nanotechnology, various kinds of academic research have been carried out concurrently and successfully along the academic progress of multi-phase flow, but would have to be continued in the future.

1.3.2 Ultrapure Water Production System

In the 1980s, tremendous progress was made in the ultrapure water production system for realizing semiconductor technologies. To improve the yield rate of semiconductor devices, the densities of contaminated solid particles and organic contaminants in the treated water are the key factor and the methods of diminishing contaminated solid particles and organic contaminants have been continuously realized. The diminishing methods would be composed of ion exchange resins, UV lights, membranes, and filters. Before establishing the ultrapure water production system, the difference between the contaminated solid particles and the ultrafine bubbles would not be clear. The number density of contaminated solid particles would limit the range of error for the generation number density of ultrafine bubbles. Therefore, the establishment of the ultrapure water production system has initiated academic and experimental approaches for the clarification of ultrafine bubbles. Figure 1.1 shows the relationship between the number density of ultrafine bubbles and the contaminated solid particles and the density of organic contaminants in water. The detection of a small amount of generated ultrafine bubbles is possible only with the usage of ultrapure water. From the viewpoints of history, it is possible to say that the realization of ultrapure water would have cleared the existence of ultrafine bubbles in the small number density region. Also with the improvement in ultrafine bubble generators, the realized maximum number density of ultrafine bubbles has become much larger than the contamination density of solids contained in city water (tap water); therefore, the number density of ultrafine bubbles for tap water has become a very important index from the viewpoints of application.

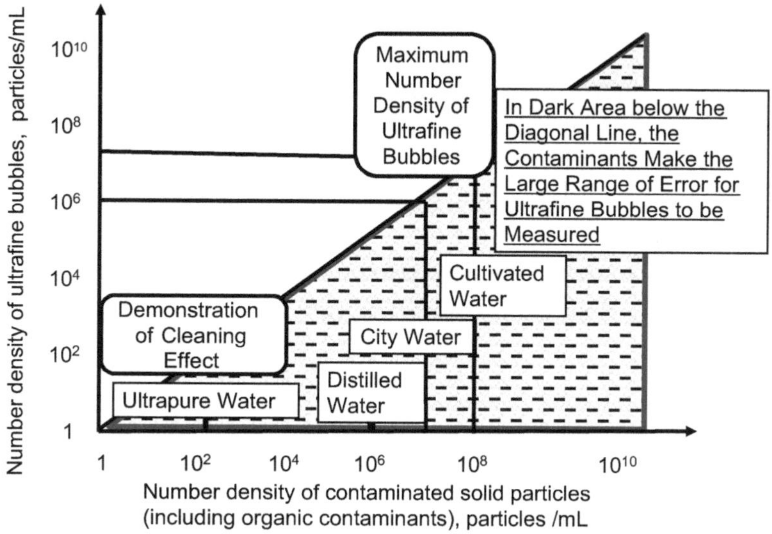

Figure 1.1 Relationship between the number density of ultrafine bubbles and the number density of contaminated solid particles (including organic contaminants).

1.4 Various Challenges for Characteristic Clarification and Possible Application of Ultrafine Bubbles

There were several challenges for clarifying the characteristics of ultrafine bubbles and investigating the applications of ultrafine bubbles from about 2000 to 2008 mainly in Japan for the first stage. Then several companies faced many challenges continuously in many application fields. In this section, various challenges will be briefly explained.

1.4.1 First Measurement of Ultrafine Bubbles in Water

By utilizing ultrapure water and by minimizing contaminated solid particles in the water, ultrafine bubbles generated by supersonic wave cavitation in water supersaturated with air have been measured for the first time [11]. The measuring instruments were particle counters for ultrapure water. The measured values of particles and bubbles increased for several tens and hundreds per mL after the initiation of supersonic wave cavitation and decreased after the termination of the cavitation by passing through the membrane filters. The diameters of the measured ultrafine bubbles are in the range of 100–150 nm. Furthermore, the amount of produced ultrafine bubbles has been revealed to depend on the dissolved oxygen concentration. The generation method of ultrafine bubbles by supersonic wave cavitation has been patented [12].

1.4.2 Cleaning Effect of Ultrafine Bubbles for Minute Particle Contamination on Plate

Since ultrafine bubbles have diameter below 1 μm, the air contained inside the bubbles is compressed by the surface tension and the pressure inside the bubbles is estimated to be larger than the atmospheric pressure. Therefore, by collapsing the ultrafine bubbles on the surface and by generating a pressure wave, the minute solid particles on the plate are estimated to be removed from the surface.

Experimental research has been conducted to verify the cleaning effects of ultrafine bubbles. Consequently, minute particles having a diameter of about 50 nm and contained on the SiO_2 wafer have been successfully removed from the wafer surface by impinging a jet of ultrapure water containing ultrafine bubbles. By impinging an ultrafine-bubbles-contained water jet for tens of minutes, it has been revealed that 98.9% of particles can be successfully removed in volume [11]. Figure 1.2 shows the photograph of a wafer in the process of cleaning by impinging a jet of ultrapure water containing ultrafine bubbles. The cleaning process of particle contaminants included circular areas from where particles have been removed. Utilization methods and apparatus of ultrafine bubbles have been systematically discussed, and various kinds of applications have been patented [13].

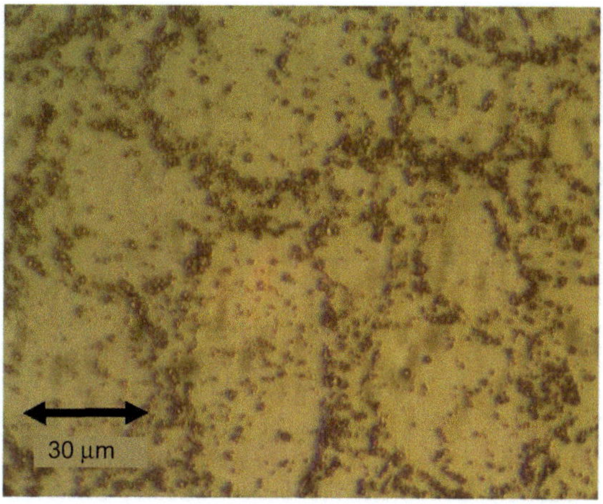

Figure 1.2 Photograph of a wafer contaminated with minute solid particles in the process of cleaning by impinging a jet of ultrapure water containing ultrafine bubbles (Yabe, A.).

1.4.3 Shrinking Microbubbles and Ozone Microbubbles

During the rising process of microbubbles that have a diameter below 50 μm, the shrinking phenomena of microbubbles occur in some cases and the generation of ultrafine bubbles is estimated [14]. By the use of ozone microbubbles, the purification of water will become more effective with the smallest amount of ozone consumption. Various applications of ozone microbubbles have been tried, and the actual proof tests have been made for oyster farming. With the addition of some amount of electrolyte, the microbubbles are assumed to become bubbles having smaller diameter sizes below 1 μm [15].

1.4.4 Transport of Solar Cell Wafers, Volume Reduction of Jellyfish Disposal, Vegetable Cultivation, and Purification of Soil

The generation devices of microbubbles have been developed since 1990 by several companies, and some devices generate not only microbubbles but also ultrafine bubbles, whose diameters have been measured mainly between 100 nm and 200 nm and have been measured for the first time in 2014 [16]. The combination of microbubbles and ultrafine bubbles has been utilized to promote the transport of solar cell wafers and the volume reduction of disposed jellyfish accumulated at the power plant cooling seawater system. Furthermore, strawberry plants have been promoted to become higher quality as an example of successful application of vegetable cultivation and purification of soil has been achieved by liquid jets containing microbubbles and ultrafine bubbles. As an application in the field of food manufacturing, sansho pepper flavor has been introduced to water by using ultrafine bubbles containing sansho pepper flavor [17].

1.4.5 Cleaning of Toilets in Expressway Service Stations and Cleaning of Salt-Stained Bridges

In 2011, cleaning of toilets in expressway service stations and parking areas by utilizing ultrafine bubbles was started in the western part of Japan (NEXCO WEST). The cleaning method of toilets and floors in restrooms utilizing ultrafine bubbles involved spraying of water containing ultrafine bubbles on the floor and toilets and wiping the floor with a mop. Before the introduction of ultrafine bubble technology, floors and toilets used to be cleaned by scrubbing with brushes and water and by absorbing the washing water with a vacuum. Therefore, the total amount of water consumption has been decreased by 1/100 and the usage of detergent has been decreased by one-third and the necessary cleaning time has been saved by 40%. Due to these tremendous effects, the cleaning method of toilets and restrooms utilizing ultrafine bubbles has spread, and about 90% of the western part expressway restrooms utilized the ultrafine bubble technology in 2018 [18]. The bridges along the expressway are sprinkled with salt in the winter for dissolving the ice on the surface of the road. Therefore, in the spring, the bridges should be washed to remove the salt from the surface of the bridges for preventing deterioration due to rust. For removing the salt from the bridge surface, ultrafine bubbles have been successfully utilized and the necessary time for removing salt from the surface has shortened to less than one-fourth with the usage of a water jet containing ultrafine bubbles [19].

1.4.6 Cleaning and Sanitizing of Vegetables, Enhancement of Vegetable Growth, Promotion of Seed Germination

Vegetables washed with water containing fine bubbles become more resistant to food-borne pathogens than vegetables washed with normal water, which has been discovered in Thailand. Using fine bubble technology for vegetables would reduce the microbial load on the surface of the incoming produce or residual pesticide, which would impact the product's quality, shelf life, and safety [18, 20]. The growth of hydroponically grown lettuce by using water containing ultrafine bubbles has been experimentally enhanced in a large-scale Japanese national project. By measuring more than a thousand lettuce stems, 20% enhancement in the weight of lettuce has been obtained after 5 weeks, which is the normal period for harvesting [21]. The promotion of germination of barley seeds has been experimentally conducted to prove 46% enhancement in the germination rate, which means the necessary time is shortened from 73 h to 39 h for 50% germination [22].

1.5 Systematic Scope of Historically Challenged Application Fields and Historical Progress Relating to Two Major Discussion Points by Academic Viewpoints

1.5.1 Systematic Scope of Historically Challenged Application Fields

As for the application of ultrafine bubbles, various kinds of challenges have been faced historically. The systematic categorization of challenged application fields will be shown from the viewpoints of application fields and the realized effects of ultrafine bubbles in Table 1.1. The application fields are categorized as engineering field, environmental field, agricultural and food field, and medical, living, and cosmetic fields. The effective functions generated by ultrafine bubbles are categorized as cleaning effect, water treatment, sterilizing promotion effect, growth promotion,

Table 1.1 Systematic categorization of challenged application fields and effective functions of ultrafine bubbles (Yabe, A.)

Systematic view of "application of fine bubble technology"								
		Effective functions						
		A	B	C	D	E	F	G
	Application fields	Cleaning effect	Water treatment	Sterilizing promotion effect	Growth promotion	Lubrication effect	Control of chemical reaction	Quality control of food
1	Engineering Application	○ [11, 18, 19, 24]	○ [29]	○		○ [23]	○ [14, 23]	
2	Environmental Application	○ [13]	○ [15]	○	○		○ [14]	
3	Agricultural and Food Application	○ [13, 18, 20]	○	○ [18, 20]	○ [21, 22]			○ [17]
4	Medical, Living and Cosmetic Application	○ [13, 26–29]		○ [25]				

○: Application technology existed (researched or developed or actually utilized or commercialized).

lubrication effect, control of chemical reaction, and quality control of food. Various kinds of application technologies that already exist will be symbolled by circles. They have been already researched, developed, utilized, or commercialized. Some literatures have also been shown in the table as reference. As the typical and successful challenges of applications, the following kinds of ultrafine bubble technology have been researched and developed so far. Application in the higher performance of grinding has been researched [23] for the manufacturing industry. The cleaning of manufacturing tools has been developed [24]. Sterilizing equipment have been researched and developed for medical applications [25]. As for washing machines, the first commercial equipment has been developed utilizing ultrafine bubbles [26]. For realizing a higher quality of living, bath and shower equipment have been developed and widely commercialized. For utilizing ultrafine bubbles and microbubbles in bathing, the effect of rise in temperature of the body surface has been observed [27]. For living applications, the moisture content of the homy layer has been observed to increase by about 20% [28]. For environmental applications, water treatment will have a large amount of application targets with simple ultrafine bubble generators [29].

1.5.2 Two Major Discussion Points of Ultrafine Bubbles

From the academic point of view, two major points about ultrafine bubbles have to be discussed. One of them is the evidence on the existence of ultrafine bubbles having a diameter below 1 μm. Another discussion point is the necessity of the theoretical mechanism explaining the long time duration of ultrafine bubbles in water. Since the inner pressure of smaller bubbles with a diameter below 1 μm is higher, the inner gas is dissolved in the surrounding water in tens of milliseconds by assuming equilibrium condition.

1.5.2.1 Evidence on the existence of ultrafine bubbles

During the past 10 years, progress in the measurement technology for the existence of ultrafine bubbles has been realized largely to show the evidence of ultrafine bubbles by two methods. The first evidence has been obtained by the resonant mass measurement method utilizing MEMS technology [30]. Cubic materials having a lighter density than water have been measured, and the materials have been observed for a long time over several months. It was estimated that the cubic materials with lighter density were ultrafine bubbles. The second evidence has been obtained from the figures of electron microscope, which are vacuum spheres with a diameter of around 100 nm obtained by the rapid freezing of ultrafine-bubbles-contained water. By analyzing the many figures, the order of 10^{10} particles/mL has been estimated for the number density of ultrafine bubbles [31]. Detailed discussions are given in the later chapters.

1.5.2.2 Hypothesis for the mechanism explaining the long-time existence of ultrafine bubbles in water

A theoretical explanation has been provided by the dynamic equilibrium model [32]. The dynamic equilibrium model assumes the adhesion of a hydrophobic material on some part of the ultrafine bubble surface. The gas inside the bubble ejects across the surface of the ultrafine bubble. But the gas is injected into the ultrafine bubble across the very tiny surface area surrounding the hydrophobic material. Theoretical calculations estimated the possibility of the long-time existence of an ultrafine bubble having a diameter less than 150 nm. The experimental evidence of this dynamic model has been obtained by using transmission electron microscopy (TEM) [33]. The state of the ultrafine bubble has been observed by the dynamic light-scattering method of TEM. Adhered organic materials have been observed on the surface of ultrafine bubbles. It can be concluded that the adhesion of organic materials inhibits the dissolution of gas from the bubbles to realize the stabilization of ultrafine bubbles. Therefore, this dynamic model is a possible mechanism of the long-time existence of ultrafine bubbles. Detailed discussions would be provided in the later chapters.

1.6 Conclusion

This chapter provides a brief history of ultrafine bubbles. Since the 20th century, the froth flotation process has been utilized in mining. Fine bubbles having a diameter of 2 μm have been used for ultrasonic imaging inside body. Microbubbles having a diameter of about 20 μm have been utilized, which became popular in the

1990s for various applications such as purification of contaminated water and growth of living beings in oceans. The challenges to wider applications have been discussed, including the shrinking phenomenon of microbubbles.

As for the historical background of the fine bubble technology, progress has been concurrently made in the academic research on two-phase flow. Also ultrapure water is essential for semiconductor devices, and the realization of ultrapure water has activated research on ultrafine bubbles without the existence of a large number of contaminated particles.

The characteristics and applications of ultrafine bubbles have been widely and intensively studied. The existence of ultrafine bubbles has been realized and measured by several methods, and the useful application effects such as the cleaning effect have been investigated and observed.

The systematic scope of historically challenged applications fields and effective functions of ultrafine bubbles has been estimated, and the fruitful future applications have been studied.

From the historical viewpoint, two major discussion points are important: the evidence of the existence of invisible ultrafine bubbles and the theoretical explanation of the mechanism for the long-time stabilization of ultrafine bubbles. The progress in the measurement technology of ultrafine bubbles has provided evidence of the existence of ultrafine bubbles, and the challenges to the theoretical approach using the dynamic equilibrium model have shown some possible mechanisms.

As for the ultrafine bubble technology, the possibility of wider applications and new effective functions is expected in the future. The investigated knowledge obtained so far will be explained in the following chapters.

References

1. Ahmed, N. and Jameson, G. J. (1985). The effect of bubble size on the rate of flotation of fine particles, *Int. J. Miner. Process.*, **14**, 3, pp. 195–215.
2. Unger, E. C. (1992). Method for providing localized therapeutic heat to biological tissues and fluids, US Patent 5149319.

3. Tachibana, K. and Tachibana, S. (1995). Albumin microbubble echo-contrast material as an enhancer for ultrasound accelerated thrombolysis, *Circulation*, **92**, pp. 1148–1150.
4. Ohnari, H. (2002). Role of microbubble technologies for multiphase flow, *Multiphase Flow*, **16**, 2, pp. 130–137 (in Japanese).
5. Ohnari, H. (2007). The characteristics and possibilities of micro bubble technology, *J. Min. Mater. Process. Inst. Jpn.*, **123**, pp. 89–96 (in Japanese).
6. Akagawa, K. (1989). A history of gas-liquid two-phase flow research and the related technology (3rd report), *Jpn. J. Multiphase Flow*, **3**, 1, pp. 2–20.
7. Clift, R., Grace, J. R., and Weber, M. E. (1978 and 2005). *Bubbles, Drops and Particles*, Academic Press and Courier Corporation.
8. Grace, J. R. (1973). Shapes and velocities of bubbles rising in infinite liquid, *Trans. Inst. Chem. Eng.*, **51**, pp. 116–120.
9. Terasaka, K. and Tsuge, H. (1993). Bubble formation under constant-flow conditions, *Chem. Eng. Sci.*, **48**, pp. 3417–3422.
10. Terasaka, K., Himuro, S., Ando, K., and Hata, T. (2016). Introduction of fine bubble science and technology (edited by the Union of Fine Bubble Scientists and Engineers), *Nikkan Kogyo Shimbun*, p. 19.
11. Morimatsu, T., Goto, M., Kohno, M., and Yabe, A. (2004). Cleaning effect of nano-bubbles, *Therm. Sci. Eng.*, **12**, 4, pp. 67–68.
12. Yabe, A. and Terakado, S. (2003). Generation method of nano bubbles, Japanese Patent, P2003-334548A.
13. Yabe, A. and Goto, M. (2004). Utilization methods and apparatus of nano bubbles, Japanese Patent, P2004-121962A.
14. Takahashi, N., Kawamura, T., Yamamoto, Y., Ohnari, H., Himuro, S., and Shakutsui, H. (2003). Effect of shrinking microbubble on gas hydrate formation, *J. Phys. Chem. B*, **107**, pp. 2171–2173.
15. Takahashi, M., Chiba, K., and Li, P. (2007). Formation of hydroxyl radicals by collapsing ozone microbubbles under strongly acidic conditions, *J. Phys. Chem. B*, **111**, pp. 11443–11446.
16. Kobayashi, H., Maeda, S., Kashiwa, M., and Fujita, T. (2014). Measurement of ultrafine bubbles using different types of particle size measurement instruments, *Int. Conf. Optical Particle Characterization (OPC 2014) SPIE* 9232 92320U.
17. Kashiwa, M., Fujita, T., Yamazaki, H., and Fushiki, T. (2012). Introduction of Sansho-pepper flavor to water by using nano-bubbles generator and

its application to the field of food manufacturing, *Annual Meeting of Japanese Society of Multiphase Flow* (in Japanese).

18. Gasiorowski-Denis, E. (2014). The fine bubble breakthrough, *ISO News*, May 12, 2014.
19. ISO/TS 21256-1 (2020). Fine bubble technology-Cleaning applications-Part.1: Test method for cleaning salt (NaCl)-stained surfaces.
20. Klintham, P., Tongchitpakdee, S., Clinsirikul, W., and Mahakarnchanakul, W. (2017). Combination of microbubbles with oxidizing sanitizers to eliminate *Escherichia coli* and *Salmonella typhimurium* on Thai leafy vegetables, *Food Control*, **77**, pp. 260–269.
21. ISO/TS 23016-1 (2019). Fine bubble technology: Agricultural applications Part 1: Test method for evaluating the growth promotion of hydroponically grown lettuce.
22. Liu, S., Oshita, S., Makino, Y., Wang, Q., Kawagoe, Y., and Uchida, T. (2016). Oxidative capacity of nanobubbles and its effect on seed germination, *ACS Sustain. Chem. Eng.*, **4**, pp. 1347–1353.
23. Kobayashi, H., Kamijyo, H., Hirano, M., and Araki, K. (2020). Application of ultrafine bubbles generation mechanism to the coolant and the realization of the higher performance of grinding, *JSPS Spring Meeting*, **141**, pp. 717–718 (in Japanese).
24. ISO 21256-2 (2020). Fine bubble technology: Cleaning applications Part 2: Test method for cleaning machine-oil stained surfaces of machined parts.
25. Yamato Scientific Co. Ltd. (2016). Development of sterilization equipment by utilizing fine bubble technology, *MEDIC (Medical Device Incubate Platform)*, H27-056, class II.
26. Washing machine utilizing ultrafine bubbles, Toshiba Zaboon AW-10M7(W), available at www.toshiba-lifestyle.co.jp/living/laundries/feature/ultrafinebubble/ (in Japanese).
27. Bath utilizing fine bubble technology and shower head utilizing ultrafine bubbles, Utility model of Japan 3148050, available at https://i-feel-science.com (in Japanese).
28. Shower head utilizing fine bubble technology, available at www.refa.net/item/refa_fine_bubble.
29. Matsunaga, D. (2012). Characteristics of microbubble generation nozzle utilizing loop flow and its applications, *Autumn Meeting of Japanese Society for Multiphase Flow*.
30. Kobayashi, H., Meda, S., Kashiwa, M., and Fujita, T. (2014). Measurement and identification of ultrafine bubbles by resonant mass measurement

method, *Proc. SPIE 9232 Int. Conf. Opt. Particle Characteristics* (OPC 2014) 92320S.

31. Uchida, T., Nishikawa, H., Sakurai, N., Asano, M., and Noda, N. (2018). Ultrafine bubble distribution in a plant factory observed by transmission electron microscope with a freeze-fracture replica technique, *Nanomaterial,* **8**, 152, pp. 1–12.
32. Yasui, K., Tuziuti, T., Kanematsu, W., and Kato, K. (2016). Dynamic equilibrium model for a bulk nanobubble and a microbubble partly covered with hydrophobic material, *Langumuir,* DOI: 10.1021/5B04703.
33. Sugano, K., Miyoshi, Y., and Inazato, S. (2017). Study of ultrafine bubble stabilization by organic material adhesion, *Jpn. J. Multiphase Flow,* **31**, 3, pp. 299–306.

Chapter 2

Introduction to Experiments

Koichi Terasaka
Department of Applied Chemistry, Keio University,
3-14-1 Hiyoshi, Kohoku-ku, Yokohama, Kanagawa 223-8522, Japan
terasaka@applc.keio.ac.jp

In this chapter, the fundamental properties of fine bubbles are described. Fine bubbles are classified into microbubbles and ultrafine bubbles, each of which has unique properties. The properties have been used in various applications that have been introduced in subsequent chapters. Next, several production methods for microbubbles and ultrafine bubbles have been introduced. Although the generation principle of microbubbles has almost been established, the generation and stabilization principles of ultrafine bubbles are still being discussed; however, manufacturing technology using these principles understood at present has been developed.

This chapter also introduces the measurement technology of ultrafine bubbles. Ultrafine bubbles are difficult to measure because they are invisible; therefore, the measurement technology is very important to evaluate and know characteristics, functions, and everything else.

Ultrafine Bubbles
Edited by Koichi Terasaka, Kyuichi Yasui, Wataru Kanematsu, and Nobuhiro Aya

ISBN 978-981-4877-59-6 (Hardcover), 978-1-003-14195-2 (eBook)
www.jennystanford.com

2.1 Introduction

Bubbles are very familiar to us. We look at them in oceans, rivers, fountains, and everywhere. There are countless types of products containing bubbles such as beverages, detergents, porous materials, and others due to their diverse properties. In addition, bubbles have been applied to technologies such as gas absorption, flotation, microbial cultivation, culturing, power generation, washing, and other technologies. Among all the attracted phenomena and science and technology, bubbles are rarely the final products or objects themselves, and they have usually been of great interest to the results of the contribution of bubble formation and presence. Thus, systematic classification covering all "bubbles" has not been completed yet.

Therefore, the definition of "bubbles" has been reconsidered. As shown in Fig. 2.1, in a broad sense, "bubbles" are "closed spaces made up of gas surrounded by the material other than gas." When classified by the phase in contact with the bubble surface, we have the following:

- "Floating bubbles," which are surrounded by liquid.
- "Partially adhering bubbles," which are in partial contact with solid.
- "Hollow particles," which are surrounded by a solid film.
- "Void," which is surrounded by solid as a continuous phase.

Furthermore, for each classified bubble, the bubble's name may be prefixed by its size, internal gas, etc.

Traditionally, bubbles were not clearly distinguished by their size. Many of the various important properties brought about by foam are more influenced by the material in and/or around the foam than the size of the foam. Furthermore, there were no simple devices or commercial generators that could artificially and easily reduce the bubble size.

However, the first commercial microbubble generator was developed in Japan before 2000, and the Japanese newspaper reported on the results of promoting the growth of oysters by spreading very small bubbles in oyster farming in Hiroshima. The name "micro bubble" was established in the Japanese society. Microbubbles are also applied in aquaculture, agriculture, clinical medicine, food industry, chemical industry, etc. Microbubble manufacturing methods have been developed and improved year by year. In addition, microbubble research is progressing academically, and many basic and applied research results have been distributed.

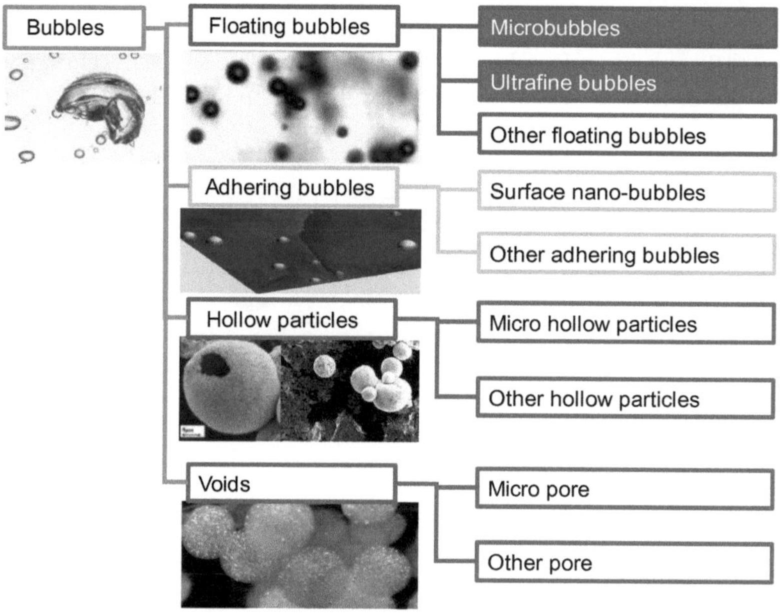

Figure 2.1 Classification of bubbles.

Furthermore, in 2007, newspapers focused on the decomposition of harmful substances by "nano bubbles," a more sophisticated microbubble. Later, the results of wastewater treatment, fishing, agriculture, and plant cultivation were also reported. Here, "micro-" and "nano-" used as prefixes of "bubble" do not mean "parts per million" or "parts per billion," respectively, but "fine" and "very fine." The meaning of a definition depends on the user or researcher, but it is ambiguous in some cases.

Due to these circumstances, it is necessary to quickly internationalize these terminology and definitions, as well as the spread and expansion of products and technologies that use "microbubbles" and "nanobubbles" in the worldwide market and the "International Organization for Standardization (ISO)." The Technical Committee of Fine Bubble Technology (TC281) was established to discuss the definition and standardization of "fine bubble." In discussions among the delegates of the ISO committee, bubbles with a sphere equivalent diameter of 100 micrometers or less were called "fine bubbles" and were distinguished from other normal bubbles [3].

In addition, they decided to call a bubble with a diameter of 1–100 μm as "micro bubble" and small bubble with a diameter of 1 μm or less as "ultrafine bubble," as shown in Fig. 2.2.

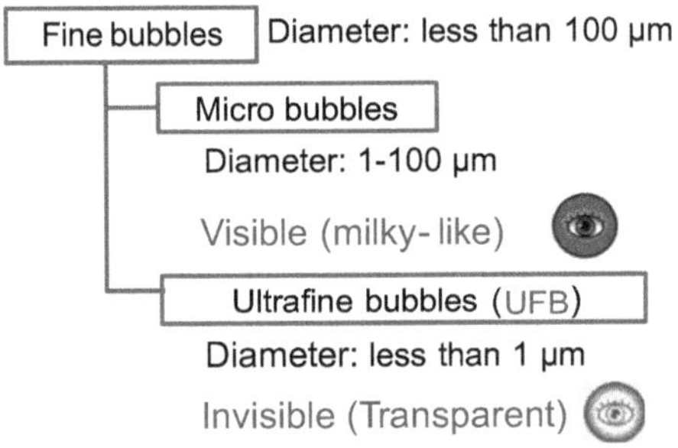

Figure 2.2 Content of fine bubbles.

In Japan, the Union of Fine Bubble Scientists and Engineers was established in 2015. Fine bubble technology has attracted attention as an innovative technology that can be applied to a wide range of fields not only in Japan but around the world [1].

2.2 Characteristics of Fine Bubbles

A comparison of typical behavior of microbubbles, ultrafine bubbles, and non-fine bubbles in water is shown in Fig. 2.3. Common bubbles that enclose air in water are non-fine bubbles, which we find in everyday life. The equivalent spherical diameter of the bubbles is 100 microns or more. Non-fine bubbles cannot maintain a true spherical shape in water but deform into an elliptical shape or an umbrella shape and rise up with a zigzag motion or spiral motion without rising straight even in static water [2].

These non-fine bubbles eventually rise to the free water surface and then rupture, and the contained gas mixes with the atmosphere above the water surface.

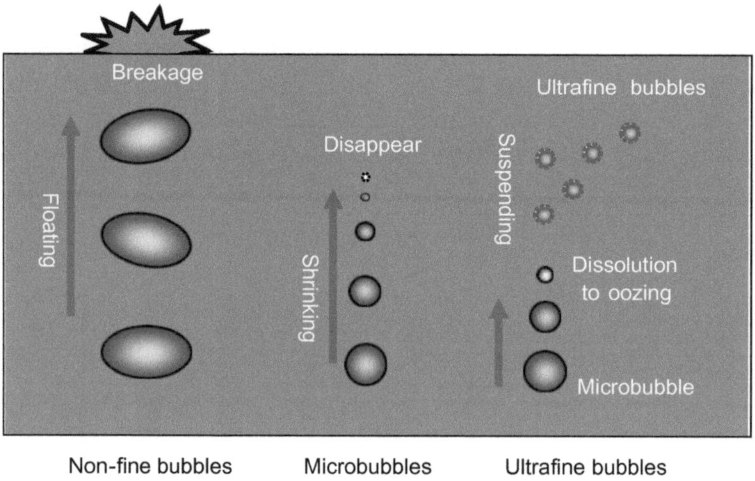

Figure 2.3 Comparison of behavior in water between fine bubbles and non-fine bubbles.

On the other hand, when microbubbles with a diameter of 1–100 microns containing air rise in water, the rising velocity is sufficiently slow compared to non-fine bubbles. Microbubbles that rise in clean water in which dissolved gas concentration is sufficiently low rapidly dissolve the contained air components into water while in contact with water, causing fast shrinkage.

On the other hand, as a characteristic behavior of ultrafine bubbles, it does not rise like non-fine bubbles and microbubbles. Therefore, no separation from the free surface due to flotation occurs. Also, once generated in water and quasi-equilibrium, it continues to be stable over a time span of several months in the case of an environment without external stimuli.

2.2.1 Rising Velocity and Brownian Motion Velocity

Figure 2.4 shows the relationship between the rising velocity u_B and the diameter d_B of microbubbles in water. The rising velocity of microbubbles can be well explained by the following Stokes equation [20]:

$$u_B = \frac{d_B^2(\rho_L - \rho_G)g}{18\mu_L} \tag{2.1}$$

where ρ_L and ρ_G are liquid density and gas density, μ_L is dynamic viscosity of liquid, and g is gravitational acceleration. The Stokes equation well represents the settling motion of a solid sphere in water under the condition of a small Reynolds number, *Re*. The relationship between the rising velocity and the diameter is in good agreement.

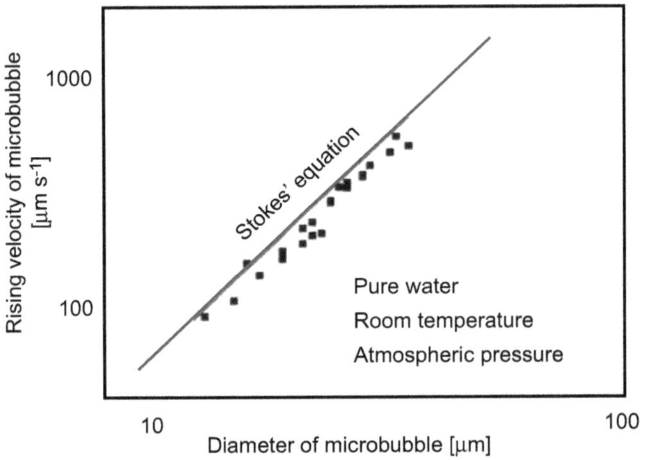

Figure 2.4 Rising velocity of microbubbles and prediction with Stokes equation.

2.2.2 Brown Movement Velocity of Ultrafine Bubble

Brown observed random motion (i.e., Brownian motion) of nano-order microparticles in a medium molecule due to collision caused by the thermal motion of media particles.

Einstein [5] derived the relationship between diffusion coefficient and particle mobility by molecular dynamics approach. For the diffusion of spherical particles at low Reynolds numbers, the Stokes resistance coefficient is used to derive the Einstein–Stokes equation (Eq. 2.2).

$$\langle \Delta x^2 + \Delta y^2 + \Delta z^2 \rangle = \frac{2T_0 k_{\mathrm{Boltz}} t_0}{\pi \mu_L d} \tag{2.2}$$

where $<\Delta x^2 + \Delta y^2 + \Delta z^2>$ is the mean square displacement [m^2] of the fine particles, T_0 is the absolute temperature [K] of the liquid, t_0 is the observation time [s], and k_{Boltz} is the Boltzmann constant [J/K].

From this equation, it can be seen that the mean square displacement is inversely proportional to the particle diameter *d*. Compared to 100 microns of a microbubble, the average square speed of 100 nm of an ultrafine bubble is estimated to be 1000 times faster.

We compared the statistical mean square displacement $<\Delta x^2 + \Delta y^2 + \Delta z^2>$ per second estimated from the Einstein–Stokes equation and the vertical rising velocity estimated by the Stokes equation.

Assume that an ultrafine bubble is a perfect spherical particle with a diameter of 100 nm and moves in water, viscosity of which is 0.98×10^{-3} Pa· s at 25°C.

Figure 2.5 shows the relationship between the rising velocity of Stokes equation (buoyancy) and the average Brownian motion velocity per second, 10 s, and 100 s of a bubble, whose diameter *d* is 10 nm to 10 μm. In the case of ultrafine bubbles ($d \leq 1000$ nm), the Einstein–Stokes random motion velocity is much higher than the Stokes rising velocity. In other words, it can be seen that the ultrafine bubbles are dominated by random motion in all directions compared to the upward motion. Due to this interesting behavior, the escape of ultrafine bubbles to the free surface is almost negligible. If there is no collision and dissolution, they remain in water for a long time.

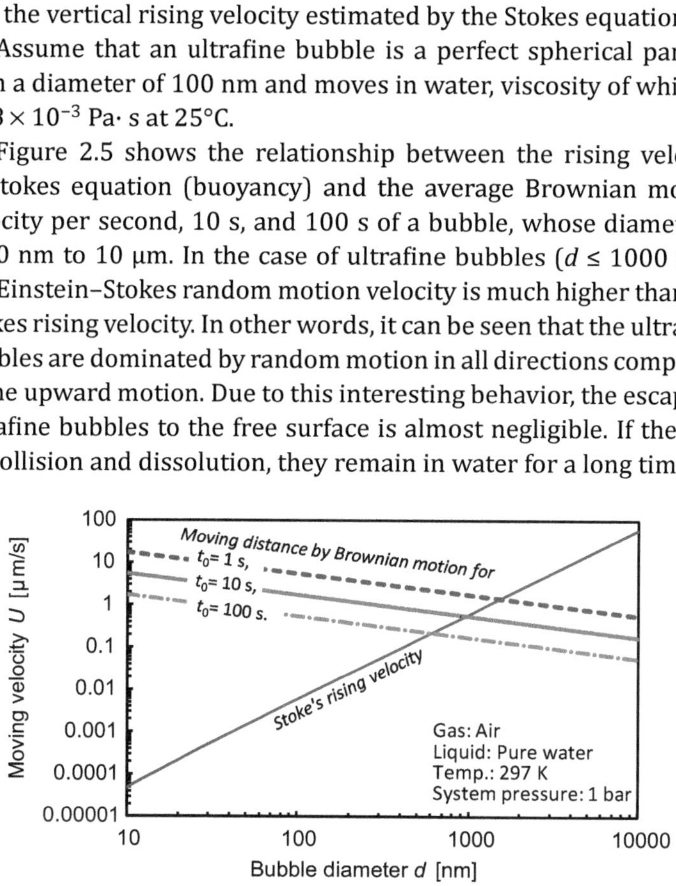

Figure 2.5 Comparison of velocity between Brownian motion and buoyancy.

2.2.3 Friction Coefficient of Microbubble Flow

Serizawa et al. measured friction coefficients for the two-phase flow when water containing 2×10^5 cm^{-3} of microbubbles flows through a circular pipe with an inner diameter of 20 mm and a length of 4 m [21]. In the case of a single-phase flow that does not contain microbubbles (α = 0%), it shows laminar flow at $Re < 2100$, and the two-phase friction coefficient decreases monotonically with increasing Re. On the other hand, when $Re > 20{,}000$, turbulent flow develops, and at the same Re, a two-phase friction coefficient is larger than that of laminar flow. For a transition region between laminar flow and turbulent flow, $2100 < Re < 20{,}000$, the two-phase friction factor decreases with increasing gas holdup (also called void fraction) of microbubble, α. Namely, it can be seen that the friction loss on inner wall is reduced. The transition from laminar flow to turbulent flow begins at about $Re = 12{,}000$ when the gas holdup (void ratio) α of the microbubble is 0.5%.

2.2.4 Specific Surface Area of Fine Bubbles

The ratio between surface area and volume of a single bubble having a diameter d_B is defined as a specific surface area. Specific ratio is expressed in Eq. 2.3.

$$\text{Specific area } a = \frac{\text{Surface area of sphere}}{\text{Volume of sphere}} \quad \frac{\pi d_B{}^2}{\pi d_B{}^3/6} = \frac{6}{d_B} \tag{2.3}$$

From Eq. 2.3, it can be seen that the specific surface area a of a bubble increases as the bubble diameter d_B decreases. When the gas holdup α is the volume of dispersed gas phase in a unit volume of aerated liquid phase, the number of bubbles n_B is expressed as follows:

$$n_B = \frac{\alpha}{\pi d_B^3/6} \tag{2.4}$$

Following Eq. 2.3, the specific gas–liquid contact area a of all bubbles floating in the unit volume liquid phase increases as

$$a = \frac{6\alpha}{d_B} \tag{2.5}$$

It can be seen that the specific gas–liquid contact area a increases as Eq. 2.5 if the same volume of gas is divided into smaller bubbles.

2.2.5 Self-Pressurization Effect of Microbubbles

As shown in Fig. 2.6, a bubble is surrounded by water or liquid molecules with large intermolecular forces. There are gas molecules in the bubble in which density is much lower than that of the liquid. The pressure inside the bubble is higher than the outer pressure by a pressure difference Δp. The pressure difference Δp can be estimated from the gas–liquid interfacial tension σ and bubble diameter d_B using the Young–Laplace equation.

$$\Delta p = \frac{4\sigma}{d_B} \tag{2.6}$$

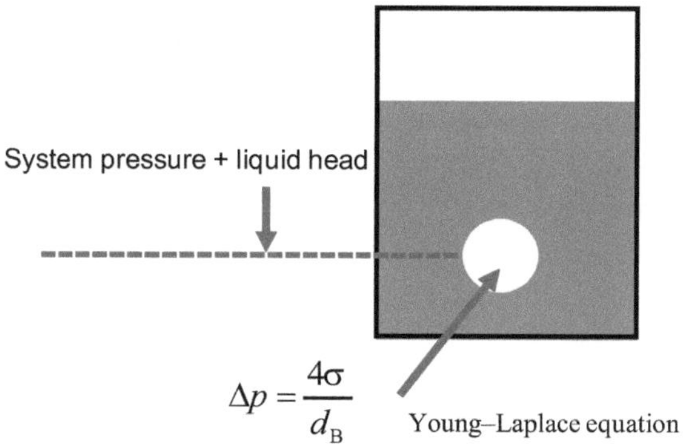

Figure 2.6 Pressure in a bubble.

Table 2.1 shows the difference between the internal and external pressures of a bubble floating in water of various diameters using the Young–Laplace equation.

Table 2.1 Internal pressure of bubbles with diameter d_B

Bubble diameter d_B	Pressure difference between inside and outside [atm]
1 mm	ca. 0.003
10 μm	ca. 0.3
1 μm	ca. 2.9
100 nm	ca. 29

When the bubble size decreases and reaches a microbubble, the pressure inside the bubble becomes very large. So it cannot be ignored compared to the atmospheric pressure (= 1 atm). The pressure on the liquid phase side is equivalent to the system pressure (atmospheric pressure) + liquid head (water pressure), but the internal pressure of the bubbles is very high, and the dissolved gas component is at a higher concentration where it is in contact with the bubble surface .

2.2.6 Microbubble Shrinkage

For microbubbles as shown in Eq. 2.3, the gas–liquid contact area per unit volume increases as the microbubble diameter d_B decreases. Figure 2.7 shows a micrograph of the microbubble shrinkage behavior and the time change in the diameter d_B of the shrinking microbubble. As the shrinkage progresses, the shrinkage rate of the microbubbles accelerates.

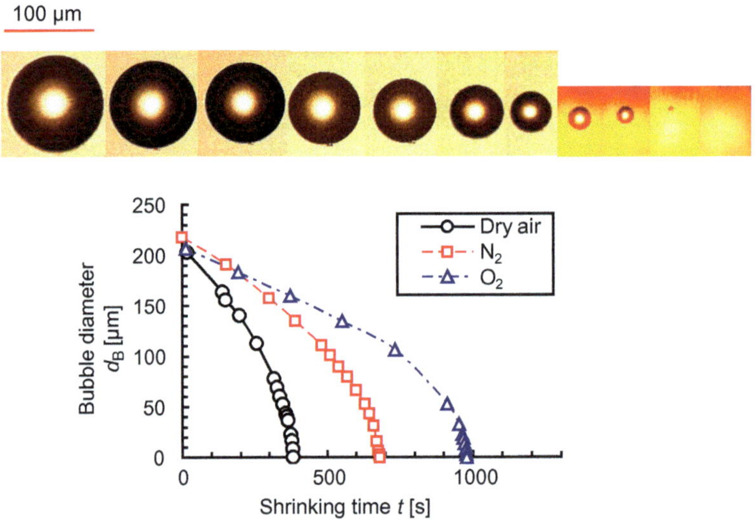

Figure 2.7 Shrinkage of microbubbles.

2.2.7 Surface Potential Characteristics of Microbubbles

It is known that the surface of a microbubble is charged just like colloidal particles in water are charged. Takahashi et al. shows a microbubble trajectory that zigzags in water under an electric field [25]. Microbubbles are guided to a container (small cell) with electrodes on both sides, and the positive and negative polarities of the electrodes on both sides are switched for about 1 s while observing bubbles in the cell with a microscope. If the bubble surface is not charged, it moves vertical upward without being affected by the electric field according to the Stokes equation introduced in Eq. 2.1. If charged, the electric field causes attraction or repulsion, and a horizontal velocity component is added to the microbubbles. The microbubbles move in a zigzag behavior in conjunction with the polarity exchange of the electrodes, and the direction of movement is always toward the positive electrode, so it can be seen that the microbubbles are negatively charged. This motion was image-analyzed, and the microbubble diameter was obtained from the moving velocity of the vertical component using the Stokes equation. Then the surface potential of the microbubble was found from the moving velocity of the horizontal component. From this result, the surface potential of the microbubbles was measured to be about –35 mV in distilled water.

2.3 Generators of Fine Bubbles

Ultrafine bubbles are usually thought to be born after the rapid shrinkage of microbubbles (some hypotheses have been discussed, although the mechanism is not yet well established).

2.3.1 Generators of Microbubbles

Microbubbles and ultrafine bubbles have so different sizes as well as properties that their manufacturing methods are also different. The technology for generating only one microbubble (with a diameter of 1 to 100 microns) in liquid is difficult even in the laboratory, and size control is particularly not easy. Furthermore, there is currently no technology capable of producing a single ultrafine bubble in liquid.

The fine bubble manufacturing method refers to a technique for mass producing microbubbles or ultrafine bubbles within a certain limited time. When microbubbles are dispersed in liquid from a microbubble generator, the microbubbles shrink and change in size depending on the state of the liquid (physical properties such as dissolved gas concentration and viscosity, mostly depending on the liquid temperature). Also, with most microbubble generation methods and devices, the time and spatial diameter distribution of the generated bubbles is not always 1–100 microns. Therefore, it is currently evaluated that microbubbles are generated when the average bubble diameter measured at a position immediately after being discharged into the liquid from the microbubble disperser is 1 to 100 microns. Basically, the principle of microbubble generation is not limited to the type of liquid or gas phase, and in practice most technologies are compatible with air–water systems. In principle, the microbubble generation technology introduced in this chapter is also water based.

This is an adaptation method for air systems. Since air is slightly less soluble in water, it can be easily dissolved and supersaturated. In order to produce microbubbles using a highly soluble gas, it is necessary to dissolve a large amount of the gas in the liquid in advance to achieve a saturation concentration. It is difficult to generate microbubbles by operating the solubility of easily soluble gases with changing temperature and pressure. On the other hand, since ions exist stably in water, electrical charging of the microbubble surface is promoted, so coalescence of microbubbles is suppressed. In addition, since the viscosity of water is small, the energy input to the water consumed when generating microbubbles is spent on crushing bubbles by shear flow rather than viscous friction loss, so more microbubbles are produced than other liquid media. However, since the interfacial tension of gas–water system is large, a large amount of energy is required to break bubbles, so it is effective to add a surfactant to lower the interfacial tension of water.

2.3.2 Microbubble Generation Technology

A number of patents have been filed as techniques for producing microbubbles. The principles can be roughly classified [27] in Table 2.2.

Table 2.2 Classification of microbubble generation principles

Principle	Types of microbubble generator
Gas crushing by liquid shear flow	Swirling liquid flow type
	Static mixer type
	Mechanical shear flow type
	Porous membrane type
Cavitation	Ejector type
	Venturi type
Gas solubility change in liquid	Pressurized dissolution type
	Heated oozing type
Phase change of dispersed phase	Mixed steam condensation type
Chemical change	Electrolytic type

Several microbubble generator manufacturers and academic research institutions have proposed and/or marketed microbubble generators. Microbubbles are manufactured using one or more of these principles. Since the principles used for generation are not identical, the generated microbubbles have different characteristics, and the generation process also is not same. The suitability and superiority of the microbubble generation principle described here vary depending on the application. Furthermore, even commercial production equipment that uses the same principle often has large performance differences and special application ranges from product to product, and it is necessary to check the specifications before purchasing or using the product.

We introduce various typical microbubble generation methods and summarize the generation principle and outline.

2.3.2.1 Swirling liquid flow type

A microbubble generator that appeared in the 1990s and was developed [17] has become popular for aquaculture of oyster in Japan. Figure 2.8 shows a schematic diagram of the device. A cylindrical generator for dispersing microbubbles in water, a liquid drive pump for introducing water with a predetermined volumetric flow rate into the generator, and a gas introduction unit with an appropriate volumetric flow rate are included. The cylindrical generator has one two-phase mixture outlet at the center of the upper end face, and a gas inlet at the center of the opposite cylindrical lower end face. At the side face close to the upper end of the cylinder is a liquid inlet through which water is fed in a tangential direction to generate a swirling water flow. The water in the cylinder does not go to the gas inlet on the lower end face but flows out toward the upper end outlet. The vicinity of the central axis of the cylinder is depressurized by the fast swirling water flow in the cylinder. When the pressure falls below the atmospheric pressure applied to the gas inlet, air is sucked from the atmosphere. The air passes through the center of the cylinder and forms a whirlpool toward the outlet of the upper end face. In the cross section inside the cylinder, it is centrifuged into a low dense air at the center and a higher density liquid on the wall surface. The gas–liquid mixture simultaneously flows while contracting while passing through the outlet hole, and then discharged into liquid bulk. At this time, a vigorous liquid shear flow is generated, and the whirlpool is crushed to generate a lot of microbubbles. The generated microbubbles are rapidly dispersed along with the water flow spread radially by the centrifugal force, so that the probability of collision and coalescence of microbubbles with each other decreases. Therefore, the average size of the microbubbles is reduced by increasing the water shear flow rate and separation rate at the gas–liquid outlet.

For the mass production of microbubbles, it is important to increase the angular velocity of the swirling water flow in the cylinder and adjust the ratio of air and water influent rate. However, the rational discharge pressure of the pump for pressing water into the cylinder and the proper design of the generator becomes important. In Fig. 2.8, water is recirculated into a microbubble generator through a suction pipeline of a pump, but a single-pass system that passes through the microbubble generator only once for both air and water is also possible. However, in the case of air–water systems, it is common to require a water flow rate that is 10 to several times that of air by volume.

Therefore, circulation systems are often designed to save the total amount of water. In addition, there are improved generators combined with submersible pumps for application in larger systems such as lakes and seas.

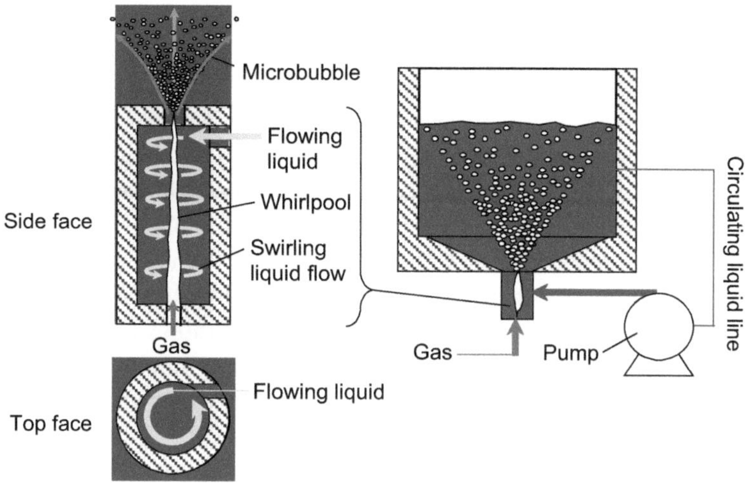

Figure 2.8 Swirling liquid flow type microbubble generator.

The number mean diameter of air microbubbles generated in water is usually about 10–60 μm. However, for other liquid than water, especially when liquid viscosity and/or gas–liquid interfacial tension are significantly different from water, the average diameter changes greatly. In the aqueous salt solution, coalescence of microbubbles is suppressed so that the number concentration of microbubbles is increased.

Furthermore, since the microbubbles are discharged in the centrifugal direction, if the distance to the wall of the dispersed bubble water container is not sufficient, the coalescence of the microbubbles is caused by a collision with the wall surface. To successfully disperse microbubbles, the container size and shape must be designed appropriately.

This type of microbubble generator has been used in a wide range of environments, including oyster farming in the bay, scallop farming, and water purification in dam reservoirs [19].

2.3.2.2 Static mixer type

The static mixer type of microbubble generator is a method in which the structure in the flow path is complicated without using a mechanical crushing operation, and a large gas bubble in liquid is crushed by a large viscous shear force mainly derived from a vortex generated by a liquid-flow driving force. Conventionally, it is used to promote liquid–liquid mixing using pipeline transportation but is often used alternatively as a gas distributor in an aeration tank. There are many different shapes and features of obstacles for exciting the turbulent vortices installed in the flow path. Figure 2.9 introduces a typical static mixer type microbubble generator [31]. In the static mixer shown in the left figure, guide plates are installed at the entrance of the microbubble generator to induce a gas–liquid swirl flow. Vortices are excited in the liquid flow by the projections on the inner wall of the cylinder, and large bubbles accompanying the liquid are crushed by shearing to generate microbubbles. The static mixer in the right figure is a type in which obstacles of a regular shape are filled in the pipeline, and the entrained bubbles are crushed using vortices when the liquid flow passes through the bent flow paths.

In addition, there are several experimental studies for developing new microbubble generation technologies. For example, methods of crushing bubbles using shear associated with a liquid jet [22] and methods of passing bubbles through a packed bed [13] are being studied.

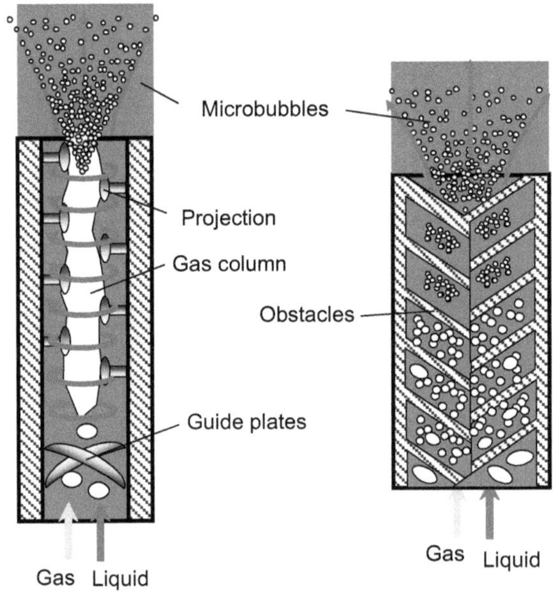

Figure 2.9 Static mixer type microbubble generators.

2.3.2.3 Mechanical shear flow type

A microbubble generation method by forced liquid flow using a motor and impeller has been developed [7]. Figure 2.10 shows the construction of a typical mechanical shear microbubble generator. Gas suction is performed at the same time as liquid suction by rotor blades, and large bubbles are crushed vigorously by the turbulent vortex of the stirring blades and the liquid generated by the stirring at the gas–liquid discharge port and then released in the centrifugal direction. Since the bubble crusher unit and the liquid drive unit are integrated, the entire system is relatively compact. The flow rate of gas and liquid introduced into the generator can be adjusted to some extent by the number of revolutions of the motor.

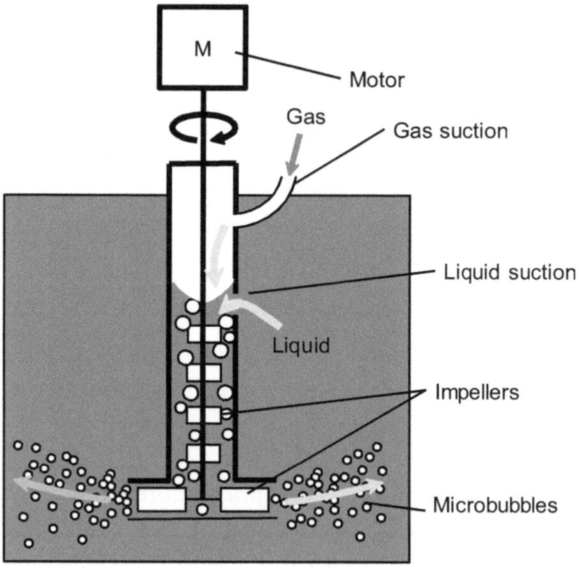

Figure 2.10 Mechanical shear flow type microbubble generators.

2.3.2.4 Porous membrane type

When gas is slowly injected from a small orifice (diameter d_O) opened in a horizontal solid plane submerged in liquid, a bubble oozes out from the orifice, and then the bubble expands up to a certain size without taking off from the orifice. As shown in Fig. 2.11, the capillary force pulls the bubble back to the orifice (this case, downward). The force acts at the contact point where gas in a bubble, liquid outside a bubble, and the solid edge of the hole. While the buoyancy (upward) of the bubble generated in proportion to the bubble volume is smaller than the capillary force, the bubble continues to expand without being released from the orifice.

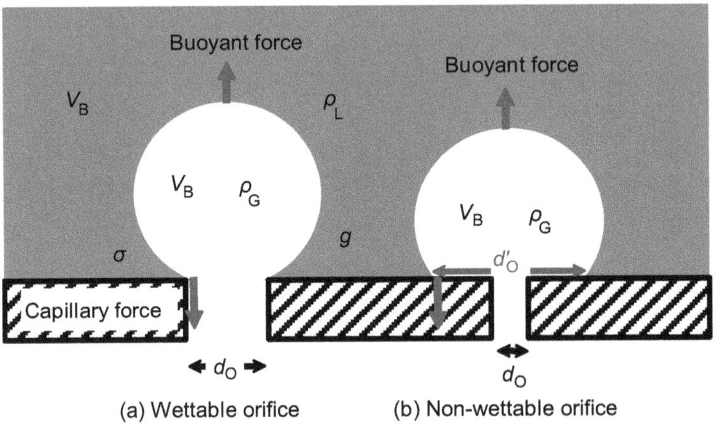

Figure 2.11 Force balance of detaching bubble from an orifice.

The force balance is described as follows:

$$(\rho_L - \rho_G) V_B g \geq \pi d_O \sigma \tag{2.7}$$

where V_B is bubble volume [m^3], σ is interfacial tension [N/m], and ρ_L and ρ_G are liquid density [kg/m^3] and gas density [kg/m^3], respectively.

The left side of Eq. 2.7 is the buoyancy acting on the bubbles, and the right side is the force (capillary force) obtained from the interfacial tension acting on the end of the orifice that prevents the bubbles from rising. As it can be seen from this inequality, the buoyancy must exceed the capillary force before the expanding bubble size exceeds the microbubble size range. For pores with good wettability as shown in Fig. 2.11a, micropores with a smaller diameter d_O are required. However, if the wettability of the solid surface with holes is poor, the gas–liquid solid three-phase contact line moves away from the periphery of the micropore of diameter d_O and spreads on the plane as shown in Fig. 2.11b. A lager bubble than a microbubble is produced with a larger diameter d_O' than the orifice diameter. This phenomenon is observed under common conditions such as a water–air system.

Therefore, in order for air bubbles that grow to be released from the orifice before exceeding the size of the microbubble, another force than buoyancy F is required as follows:

$$(\rho_L - \rho_G) V_B g + F \geq \pi d_0 \sigma \tag{2.8}$$

In the porous membrane type microbubble generator, an improvement in the material wettability as well as shear liquid flow to assist in releasing bubbles from the orifice is required.

By applying an external force F, it is possible to promote separation of bubbles from the pores while the size of the bubbles is still small.

Figure 2.12 shows the structure of a typical microporous membrane type microbubble generator. Gas is supplied to a cylinder made of a porous membrane. Bubbles grow from very small holes on the sides of the cylinder and flow into the annular path of the coaxial double cylinder. Due to the shear flow of the liquid perpendicular to the direction of bubble growth, the expanding bubbles are cut and flowed downstream with the liquid flow.

The size and number concentrations of microbubbles vary to some extent depending on the balance between gas flow and liquid flow. In particular, the material of the porous membrane, the pore diameter, and the aperture ratio are devised by each microbubble generator manufacturer, and the pore diameter and thickness of the porous membrane have a great influence on the pressure loss of the penetrating gas [10]. In addition, membrane production technology is required to generate microbubbles at a reasonable gas pressure. A microbubble dispersion method in microchannels using micropores has also been developed [34].

In addition to utilization of liquid flow, a method has been developed to quickly let bubbles detach from fine holes using ultrasonic vibration. When separation is promoted by applying ultrasonic waves to the bubbles generated in the liquid from the tip of the needle-shaped microtubule, the generated bubbles become microbubbles [11]. The generation of microbubbles has been confirmed in an ultrasonic field of several tens of kHz. The larger the system volume, the greater the input energy necessary, and so it is expected that microbubbles should be supplied to the static liquid in a small space.

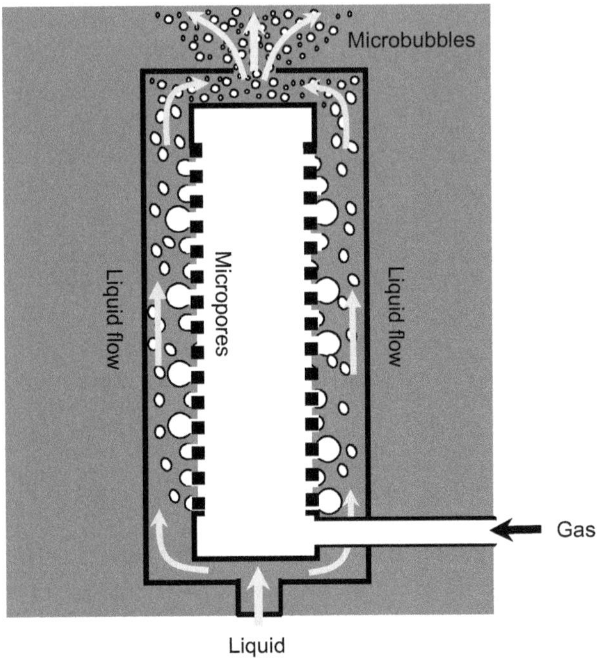

Figure 2.12 Typical porous membrane type microbubble generator.

2.3.2.5 Ejector type

In the ejector, the gas is sucked in by using the reduced pressure generated by the liquid flow passing through the narrow path at high velocity, and the sucked gas is crushed due to the cavitation caused by the expansion of the downstream pipe. Figure 2.13 shows the structure of a typical ejector type microbubble generator. Thus, there are many microbubble generators that use abrupt shrinkage and expansion of the flow path to generate reduced pressure in the pipe and then suck gas.

Ejector type microbubble generators with spherical obstacles on the pipe [18] and the other type using slits [15] have been developed. The device can be manufactured easily.

For example, it can be manufactured by modifying a device already sold as a water pump. The microbubbles size produced with the microbubble generators is relatively large, and the amount of microbubbles generated relative to the amount of liquid introduced is relatively small.

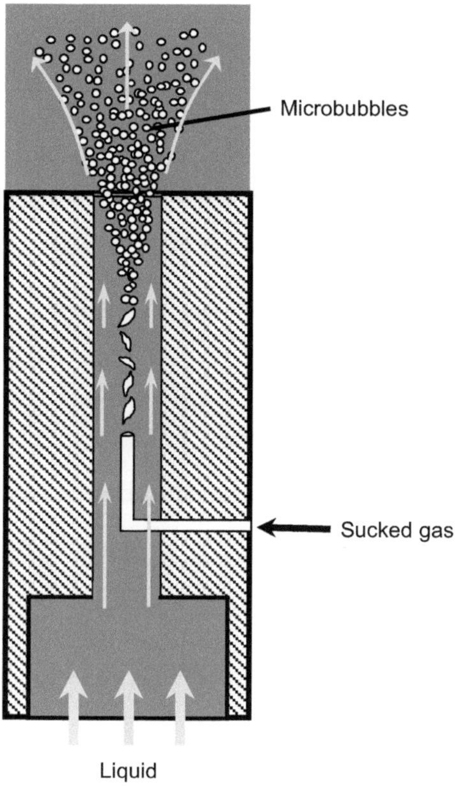

Figure 2.13 Ejector type microbubble generator.

2.3.2.6 Venturi type

When a liquid containing large bubbles is allowed to pass through a Venturi tube having a contracted and then enlarged cross-sectional area as shown in Fig. 2.14, the bubbles are once expanded by the sudden pressure reduction when passing through the throat part, and then those are suddenly crushed. The bubbles collapse violently due to the pressure recovery and shock wave [26]. Venturi type microbubble generators designed by this principle have become widespread in the industry [6]. As with the ejector type, the point of using rapid pressure reduction and pressurization in the pipeline is the same, but since self-priming of gas using the pressure reduction part is not necessarily required. Therefore, the ejector type and Venturi type generators were classified separately. The device shape of Venturi is extremely simple, and hydrodynamic analysis has been advanced by many researchers. The inner diameter of the narrowest channel can be made relatively larger than that of other microbubble generators. On the other hand, since the microbubble groups born in the Venturi tube are close to each other, collision and coalescence frequently occur and have a relatively large bubble size distribution.

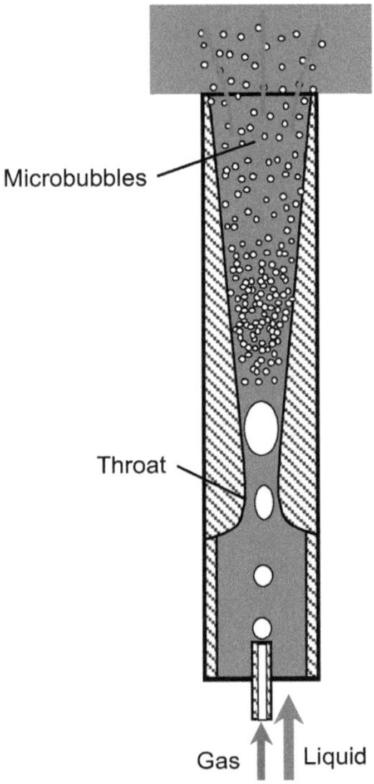

Figure 2.14 Venturi type microbubble generator.

2.3.2.7 Pressurized dissolution type

The pressurized dissolution method is one of the microbubble generation technologies that use the physicochemical pressure dependence of the solubility of gas in a liquid. Figure 2.15 shows the result of estimating the solubility of the main components of air (79% nitrogen and 21% oxygen) in water and the pressure from Henry's law.

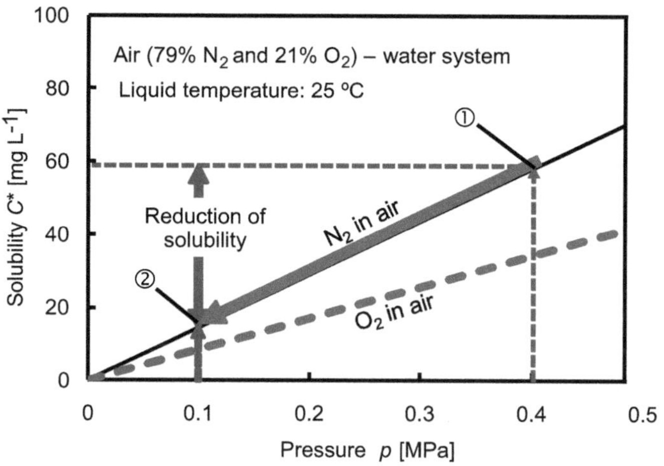

Figure 2.15 Pressure dependency on the solubilities of N_2 and O_2 gases in water.

In atmospheric pressure (0.1 MPa) and room temperature water, respectively, 79% nitrogen and 21% oxygen in the air dissolve about 15 mg/L and 9 mg/L. However, because solubility is an equilibrium value, it is necessary to be aware that the concentration may temporarily exceed the solubility (supersaturation) until the solubility is reached.

When air is pressurized in the container, the partial pressure p increases because the total pressure increases even if the composition (molar fraction) of nitrogen and oxygen in the air is the same. As shown by Henry's law, the saturation concentration of gas into water increases as the partial pressure increases. For example, when pressurized to 0.4 MPa, nitrogen and oxygen can dissolve up to about 58 mg/L and 34 mg/L, respectively. Therefore, when the water that was pressurized to 0.4 MPa in Fig. 2.15 and that reached vapor–liquid equilibrium is decompressed at a constant temperature to 0.1 MPa in ②, the difference in saturation concentration is about 43 mg/L and about 25 mg/L. Thus, nitrogen and oxygen appear as gases from water.

Figure 2.16 shows the bubble generation mechanism from gas supersaturated water. When water saturated with gas under a pressurized system is released to normal pressure, the dissolved gas saturation concentration C^* under the pressure is suddenly reduced to the saturation concentration C^* at normal pressure. The supersaturated dissolved gas does not remain dissolved in water but appears from the liquid phase to the gas phase. If this process is considered by applying the equilibrium theory, all of the supersaturated dissolved gas will ooze during a sufficiently long time, and the total gas volume can be calculated from the difference in saturation concentration. However, the process of the change is not constant depending on the conditions, so it is necessary to consider the mass transfer theory considering time.

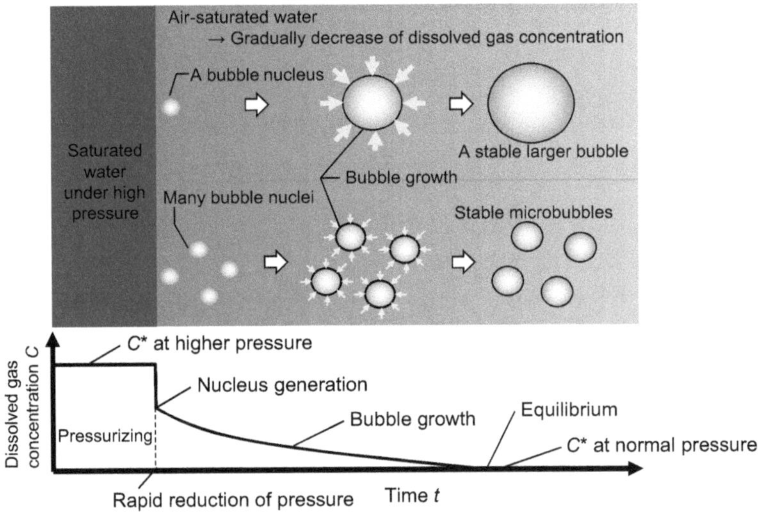

Figure 2.16 Bubble generation mechanism from supersaturated water by pressurized dissolution.

As shown in Fig. 2.16, when a sudden depressurization operation is performed and the dissolved gas saturation concentration C^* in water suddenly reduces, a certain amount of gas component is oozed from water as predicted from the gas–liquid equilibrium theory. Nucleation and bubble growth occur in a chain or simultaneously. Bubbles are born at certain points in water, which is called a bubble nucleus. When a liquid is composed of multiple components, it is hardly that each component is perfectly distributed. With a little temperature differences, history of mixing and dissolution processes, contact with different phases, contaminants, rough face of wall, liquid inhomogeneities and slight pressure distributions, slightly dense spots are caused at certain times and locations. The dense spots become bubble nuclei. When the excess enthalpy of gas component dissolved in liquid exceeds the potential for gas–liquid interface formation, a bubble growth begins. Once stable bubble nuclei are formed, the dissolved gas concentration surrounding the bubble nuclei decreases, so that the dissolved gas transfers from liquid bulk into the bubbles. Namely, the bubbles expand.

This is called bubble growth.

As learned from the transport phenomena theory, the transfer rate of gas components dissolved in liquid is governed by the diffusion coefficient or mass transfer coefficient, so bubbles may be born in different time and position. However, flowing, ascending, coalescence, and breakup of the generated bubbles are not considered. The bubble growth is completed when the concentrations of the gas phase and the liquid phase finally reach equilibrium.

Utilizing the physicochemical vapor–liquid equilibrium theory and kinetic theory, a pressurized dissolution type microbubble generation method has been developed. As described above, since the total volume of bubbles generated is constant, the more bubble nuclei, the smaller the dissolved gas components supplied per one. In order to reduce the average diameter of the microbubble swarm and increase the microbubble number density, it is necessary to increase the number density of bubble nuclei and suppress bubble growth.

Figure 2.17 shows the structure of a typical pressure dissolution type microbubble generator. Water kept in a container at normal pressure is sucked through a pipe with a pump. A back-pressure valve is placed in the suction pipe. The liquid flow path is throttled by the valve. The pressure in the pipe between the pump suction port and the back-pressure valve becomes lower than the atmospheric pressure.

Using the negative pressure, air is sucked into the pipe from the atmosphere. Both the sucked air and the water supplied from the water tank enter the pump at the same time and are mixed strongly.

The pressure is increased while the gas dissolution proceeds, and the dissolved gas saturated aqueous solution under pressure is discharged. The higher the discharge pressure, the higher the dissolved gas concentration, but it is determined by a balance with reasonable pump performance.

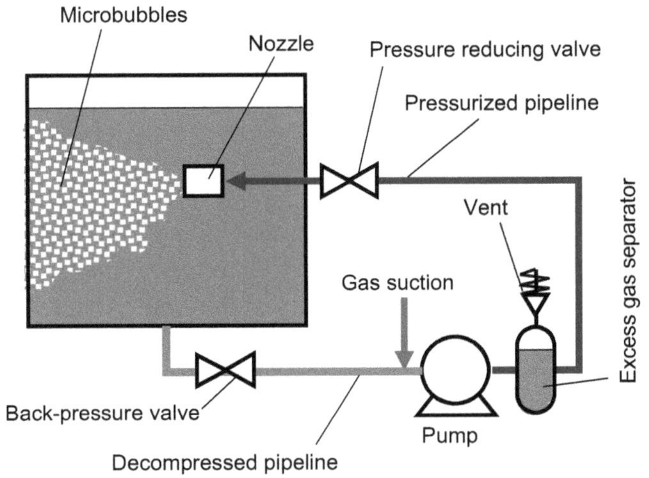

Figure 2.17 A typical system of a pressurized dissolution type microbubble generator.

A pressure-reducing valve is installed in the pump discharging pipe. The inside of the pipe from the pump discharge port to the pressure-reducing valve is kept under pressure.

In normal pumps, if gas is mixed inside, it may cause a failure or trouble. The discharge pressure must be maintained at a predetermined pressure or higher.

Water that has become supersaturated under normal pressure after passing through the pressure-reducing valve and returned to normal pressure is prompted to generate bubble nuclei when passing through the "nozzle." Since the number concentration of the bubble nuclei to be generated varies depending on various stimuli, the design and specifications of the nozzle are very important. Bubble nuclei released into the liquid bulk along with the liquid flow from the nozzle grow into a large number of microbubble swarm by absorption of supersaturated dissolved gas in water by mass transfer.

Air is rapidly dissolved in a pump by gas–liquid pressurization and agitation, but depending on the conditions, the air cannot be completely dissolved, and excess large bubbles are mixed into the pressurized water and reach the nozzle. Large bubbles cause a huge amount of dissolved gas to be sucked in and inhibit the generation of microbubbles. Therefore, a surplus gas separator for floating and separating undissolved large bubbles is placed in the discharge piping of the pump. In the separator, only water saturated with dissolved gas not containing bubbles is guided downstream, and surplus gas accumulated in the separator is discharged from the vent.

The diameter of microbubbles generated by a pressured-dissolving microbubble generator is smaller than that by a gas crushing type microbubble generator. The bubble size distribution is relatively narrow, and the number concentration of microbubbles is higher.

Therefore, the generated microbubble water is suspended in milky white. The number of microbubbles generated depends on the physical properties such as the solubility and mass transfer coefficient of the selected gas and liquid. When the gas is selected other than air and the liquid is selected other than water, it is necessary to adjust the gas–liquid ratio and suitable pressure.

2.3.2.8 Heated oozing type

Similar to the pressurized dissolution type microbubble generation method, this is a method that utilizes a change in the solubility of gas in liquid. The pressurized dissolution type microbubble generation method uses the pressure dependency of the solubility, but the heated oozing type microbubble generation method uses the temperature change in the solubility.

Figure 2.18 shows the relationship between the temperature dependency of solubility of nitrogen (79%) and oxygen (21%) in water.

The equilibrium concentrations of nitrogen and the oxygen in water at 10°C which is equilibrium with atmospheric air consisting of 79% nitrogen and 21% oxygen are approximately 19 mg/L and 11 mg/L, respectively. When 10°C water saturated with air is heated to 45°C, for example, the saturated solubility of nitrogen and oxygen in the air in water is reduced to about 12 mg/L and about 7 mg/L, respectively. Therefore, when water that has reached the vapor–liquid equilibrium at 10°C in ① of Fig. 2.18 is heated to 45°C in

②, the difference in saturation concentration is about 7 mg/L for nitrogen and about 4 mg/L for oxygen. Both gases ooze from water.

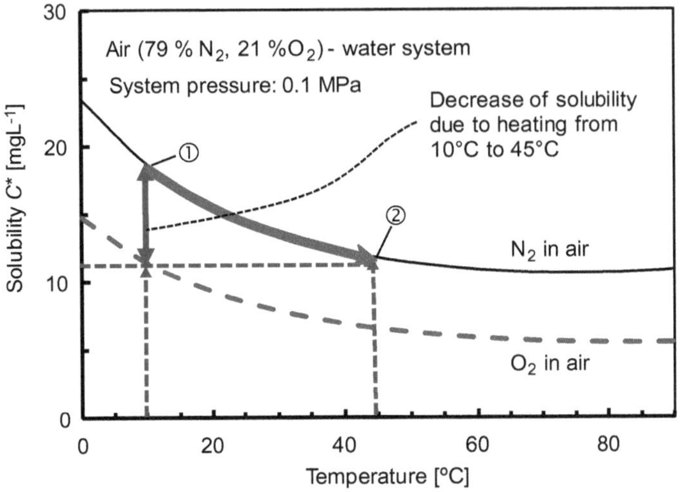

Figure 2.18 Temperature dependency on the solubilities of N_2 and O_2 gases in water.

Figure 2.19 shows the mechanism of the generation of heated oozing microbubbles. The saturated concentration of air in water decreases by heating.

Reduction in saturation causes bubble nucleation, followed by bubble growth. However, the temperature range for heating and cooling is limited because microbubbles cannot be generated if the liquid changes to ice or vapor.

The pressurized dissolution microbubble generation method can generate a large number of microbubbles by controlling the gas solubility by the relatively large pressure difference between pressure and depressurization. Whereas the heated oozing microbubble generation method generates microbubbles, the total gas played is rather small. As shown in Fig. 2.16, since the propagation of pressure change in liquid occurs at the sound speed, bubble nucleation tends to occur faster than the mass transfer rate of the dissolved gas in liquid. On the other hand, in the heated oozing type microbubble generation method, it is considered that the heat is transported in liquid by convection heat transfer to raise the temperature, and the heating rate is not significantly faster than the mass transfer rate.

Therefore, the number concentration of bubble nuclei is estimated to be relatively small.

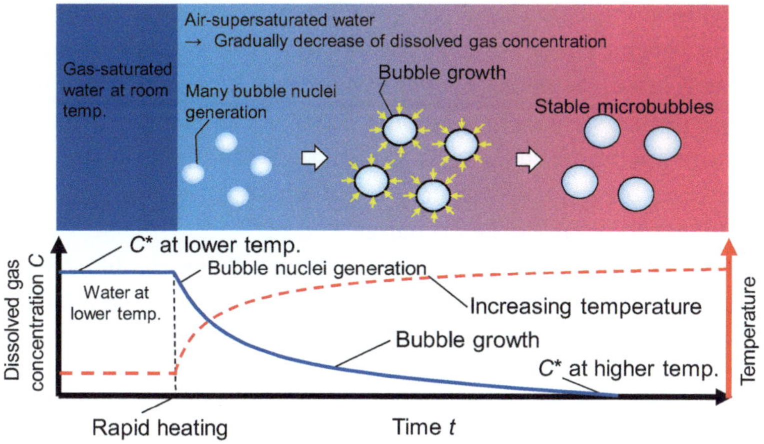

Figure 2.19 Bubble generation mechanism from supersaturated water by heated oozing.

However, depending on the system, region, and country, tap water from urban water treatment plants, which is widely used, is pressurized to about 0.25 MPa (in Japan) for the distribution pipeline and transported over long distances and is sufficiently saturated with air. For this reason, even a simple system using the heated oozing method as shown in Fig. 2.20 can deposit a milky cloud of microbubbles with sufficient turbidity in the bathtub, so that the use of microbubble generators for bathing has also spread to the market.

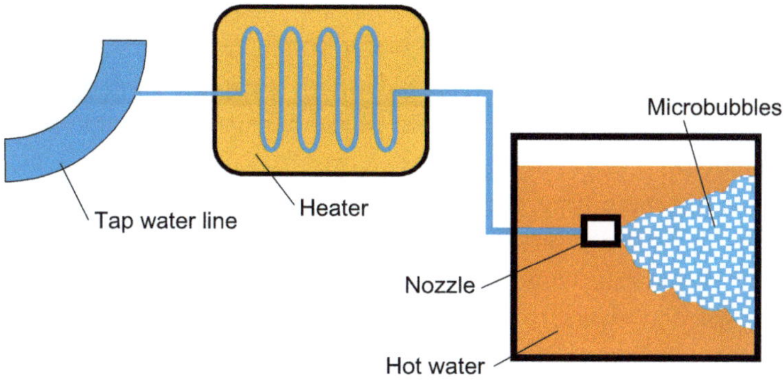

Figure 2.20 Microbubble generation system with heated oozing method.

2.3.2.9 Mixed steam condensation type

In the heated oozing type microbubble generation method, heat is used by the sensible heat change in liquid. On the other hand, in the mixed steam condensation type microbubble generation method, heat is used for phase change. Water can be easily converted to superheated steam using a boiler. Since steam is a gas if it is kept at a temperature equal to or higher than the boiling point at the pressure, other gases can be mixed with steam very simply and a homogeneous gas can be prepared.

As shown in Fig. 2.21, a small amount of non-condensable gas such as nitrogen is mixed with condensable steam. The mixed gas steam is injected into the cooling water from a nozzle, sub-milli-vapor bubbles are dispersed into water.

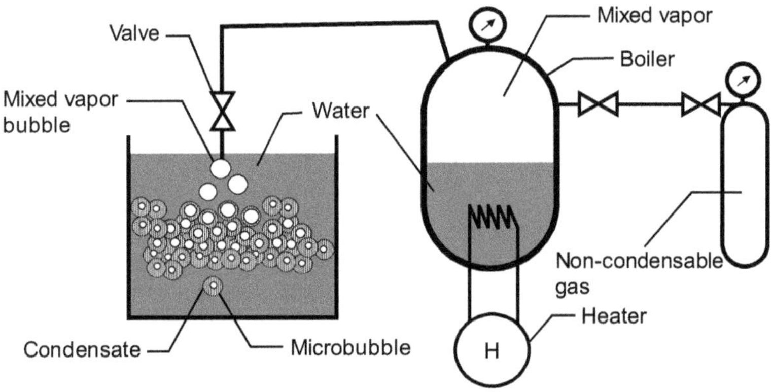

Figure 2.21 Mixed steam condensation type microbubble generation system.

As shown in Fig. 2.22, the water steam component in a mixed vapor bubble dispersed into cold water is rapidly cooled from the gas–liquid contact interface and phase changed (condensed) to be liquefied.

At the same time, the interface of the vapor bubbles is crushed violently by the momentum of the vapor plume ejected in a jet form from the nozzle while abruptly contracting, and fine bubbles are dispersed in water.

In each dispersed bubble, nitrogen that does not condense at the water temperature is maintained except for the amount dissolved in water, but the water vapor component changes into liquid water as it cools and assimilates with the liquid phase around the bubble.

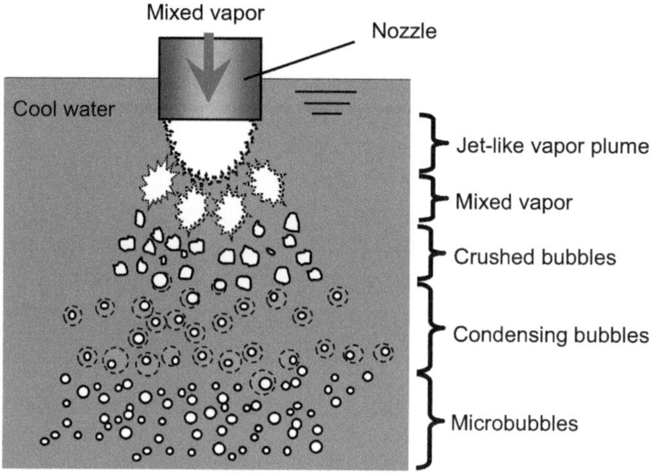

Figure 2.22 Microbubble generation with direct contact condensation of mixed steam.

As a result, the bubble shrinks rapidly, creating a swarm of microbubbles consisting only of nitrogen and water vapor with vapor–liquid equilibrium partial pressure [28].

The mean diameter of the microbubble swarm is mainly determined by the composition of steam and non-condensable gas, injection nozzle geometry, and mixed steam injection.

Almost no large bubbles are generated in this method. By using high-pressure steam laid as a general factory utility, non-condensable gas microbubbles can be produced in water. There are no mechanical moving parts such as membrane and pumps in the system.

When the steam condenses, the heat makes the water phase rapidly warm by direct contact heat transfer, so the non-condensable gas, which is dissolved at a high concentration in the liquid phase, is oozed as bubbles.

Although it cannot be used for applications where the temperature rise of liquid phase is inappropriate, it is a method that can perform high-efficiency heat transfer and microbubble dispersion at the same time when heating is desired simultaneously.

2.3.2.10 Others

As described earlier, the microbubble generation method has been introduced as a mechanical method, and a physicochemical method utilizing pressure, temperature, and phase changes, but other chemical methods are also being studied. For example, attempts have been made to produce microbubbles of oxygen and hydrogen by the electrolysis of water. In addition, a technique for producing tiny carbon dioxide bubbles by the chemical reaction of baking soda or citric acid is also used for bathing agents and detergents. Furthermore, by adding chemicals such as surfactants, the stability of the microbubbles dispersed by the microbubble generation method introduced above was improved, and the average diameter of the generated microbubbles was reduced. There are other important microbubble generators that could not be introduced here.

2.3.3 Generators of Ultrafine Bubbles

The generation of ultrafine bubbles (with a diameter of 1 micron or less) is more difficult than the generation of microbubbles. In many cases, ultrafine bubbles are born in a mixture with microbubbles. Ultrafine bubbles stay in liquid for a long time, unlike microbubbles that rise slowly. Therefore, to make water in which only ultrafine bubbles (ultrafine bubble water) are suspended, it is necessary to wait a while after fine bubble production. Therefore, the generation of most ultrafine bubble water is done batchwise.

Furthermore, the mechanism of generation of ultrafine bubbles has not yet been fully elucidated. Most methods of generating ultrafine bubbles are still limited to water–air (nitrogen or oxygen as an air component). We write down the mechanisms that include hypotheses that have not yet been fully elucidated in this book as hints for readers to understand the phenomenon.

2.3.4 Ultrafine Bubble Generation Technology

The technology to generate ultrafine bubbles is still under development, and it seems that the technology has not been completed. However, the principle of generation of ultrafine bubbles can be roughly classified in Table 2.3 [29].

Table 2.3 Classification of ultrafine bubble generation principles

Principle	Type of ultrafine bubble generators
Gas solubility change in liquid	Pressure dissolution type
Gas dispersion and mixing by strong liquid shear flow	Static mixer type
	Swirling liquid flow type
	Batch high-speed shaking method [23]
Interfacial tension drop	Membrane with surfactant type
Ultrasound	Ultrasonic cavitation method

2.3.4.1 Pressurized dissolution type

The pressurized dissolution type ultrafine bubble generation method has been evaluated as a relatively good method for microbubble generation. Figure 2.23 illustrates a typical system of ultrafine bubble generator. The water in the tank is pressurized and pumped into the pipe. Liquid flow is accelerated in the contract pipe. Dynamic pressure becomes lower than the atmospheric pressure. As a result, after the air is sucked from the atmosphere, the pipe is expanded to decelerate the air–liquid two-phase flow.

The nozzle that restricts the flow rate keeps the pressure in the pressured pipeline and dissolves the air in water to the maximum extent. The air that has not been completely dissolved in water is not sent to the nozzle and is discharged as surplus air.

The water saturated by air dissolution under pressure is decompressed to the atmospheric pressure at effluent from the nozzle, and dissolved air corresponding to the decrease in solubility is deposited as a mixture of microbubbles and ultrafine bubbles in the water tank.

This fine bubble water is pumped and circulated several times. Thereafter, when the water circulation is stopped, the microbubbles in the water tank gradually rise and leave from the water phase. The water becomes transparent, in which a lot of ultrafine bubbles having an average diameter of about 100 nm stay without rising. This is used for various purposes as “ultrafine bubble water.”

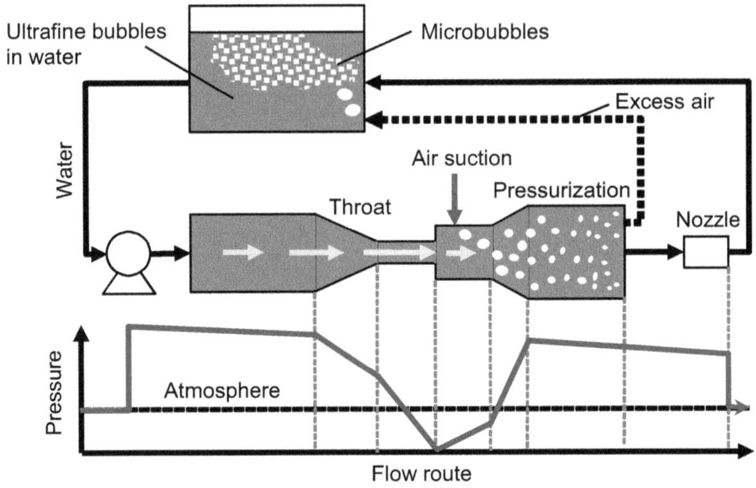

Figure 2.23 Ultrafine bubble generation with pressurized dissolution method.

2.3.4.2 Swirling liquid flow type

Figure 2.24 shows the structure of a typical swirling liquid flow type ultrafine bubble generator [32]. The swirling liquid flow type method that is widely used in a microbubble generator is improved, and a number of annular grooves are carved on the inner surface of the cylinder so as to generate a stronger and more stable swirling liquid flow. As a result, the air entrained with the water in the generator cylinder is strongly pulverized by a strong shear flow to generate ultrafine bubbles together with microbubbles. Similar to the pressurized dissolution type ultrafine bubble generator, in order to obtain a sufficient concentration of ultrafine bubbles, it is necessary to return the ultrafine bubble water discharged from the outlet to the inlet and circulate it by an appropriate number of circulations.

In this method, the dissolved gas concentration surrounding water is relatively low than that of the pressurized dissolution type ultrafine bubble generator. Since the structure of the machine is relatively simple, it is suitable for use with contaminated water such as tap water and agricultural water.

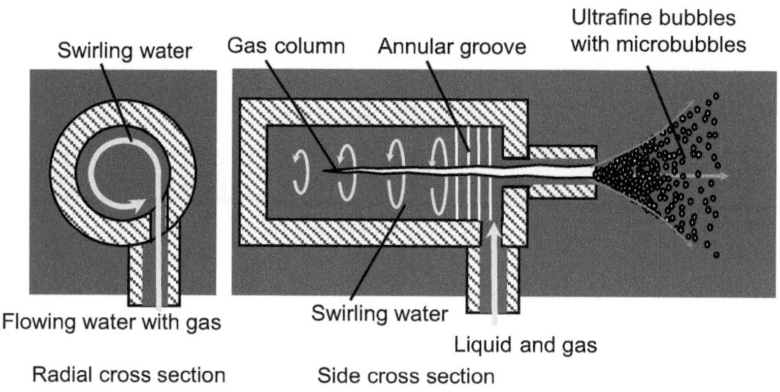

Figure 2.24 Ultrafine bubble generation with pressurized dissolution method.

2.3.4.3 Static mixer type

A static mixer type ultrafine bubble generator [14] as represented in Fig. 2.25 has been developed. A number of disk type static mixers having honeycomb-like irregularities are stacked in the pipe and inserted. Since the honeycomb-shaped irregularities of each disk do not coincide with each other when they are overlapped, they are displaced from each other, so that small cells adjacent to each other are arranged. When a liquid containing large bubbles is caused to flow in such a complicated bent flow path, a large shear flow is generated when moving from a cell to the next cell, and the large bubbles accompanying the liquid are crushed one after another.

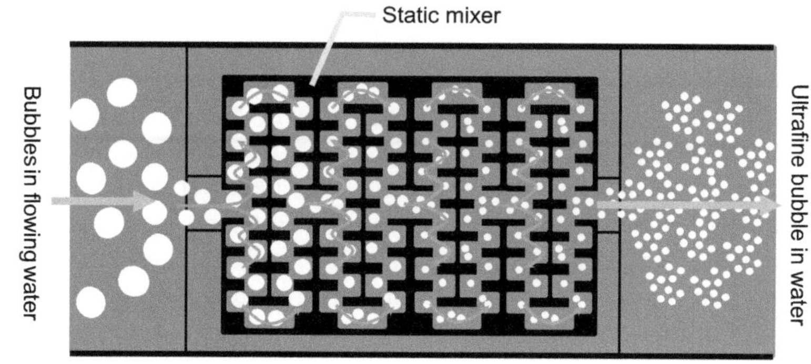

Figure 2.25 Static mixer type of ultrafine bubble generation.

Furthermore, the gas–liquid two-phase flow that has passed through a certain cell is branched into a flow to a plurality of cells, and the gas–liquid two-phase flow from a plurality of adjacent cells merges in a cell. For this reason, the folding of the liquid proceeds along with the crushing of the bubbles, and the gas dissolution into the liquid also proceeds very rapidly. When it finally passes through the static mixer, the water becomes cloudy by generating both a microbubble swarm and an ultrafine swarm, which is returned to the entrance of the static mixer and circulated for an appropriate number of times. When the cloudy water finally obtained is left standing, all the microbubbles gradually leave from the water phase, leaving a water that looks transparent to the naked eye. This type of "ultrafine bubble water" is used in aquaculture and for maintaining the freshness of fish.

The ultrafine bubble production technology introduced above has the following common points:

In any ultrafine bubble generator, water containing only the ultrafine bubble group, i.e., ultrafine bubble water, cannot be produced continuously. Microbubbles always coexist in a water tank at the generator's outlet, so if the microbubbles are not needed, we need to wait until only the ultrafine bubbles remain. Also the number concentration of ultrafine bubbles in water does not achieve so high when only one pass production apparatus.

In order to obtain ultrafine bubble water having a high number concentration, the ultrafine bubble water is usually circulated through the generator, and the number of circulations has an optimum value unique to each ultrafine bubble-producing system.

This is because the crushing and dissolution of the microbubble group further proceed by the circulation, and the increase in the dissolved gas concentration in the water contributes to the generation and stabilization of ultrafine bubbles. On the other hand, excessive liquid circulation does not necessarily increase the ultrafine bubble number concentration. The main cause is a rise in the water temperature due to heat sources such as motors in the generator. An increase in water temperature causes a decrease in the concentration of dissolved gas in the water that is the raw material for fine bubbles.

In addition, when the ultrafine bubble swarm is generated with a system consisting of clean air and ultrapure water, the operating conditions can be set using any of the shear flow type, static mixer type, and pressurized dissolution type ultrafine bubble generation methods. Even if it is changed as much as possible, a distribution of ultrafine bubbles having a single-mode diameter peak at about 100 nm is generated.

From the aforementioned comprehensive observation, it is considered that the generation mechanism of ultrafine bubbles is mainly due to the very local physicochemical behavior of gas molecules dissolved in water, not the production equipment structure or hydrodynamic causes.

2.3.4.4 Microporous membrane type with surfactant addition

Even if a reasonable external force F is applied to promote the detachment of bubbles generating from the micropores, the bubbles must have the minimum volume V_B expressed by Eq. 2.9.

Therefore, to reduce the gas–liquid interfacial tension σ by adding a surfactant, V_B can be minimized.

$$(\rho_L - \rho_G)V_B g + F \geq \pi d_0 \sigma \tag{2.9}$$

Figure 2.26 shows an example of an ultrafine bubble generator used in combination with the addition of a surfactant to liquid [10]. Usually, the normal-temperature water–air system has an interfacial tension of about 72 mN/m, but it is possible to reduce the interfacial tension σ to the minimum value by adding an appropriate surfactant to a critical micelle concentration or higher.

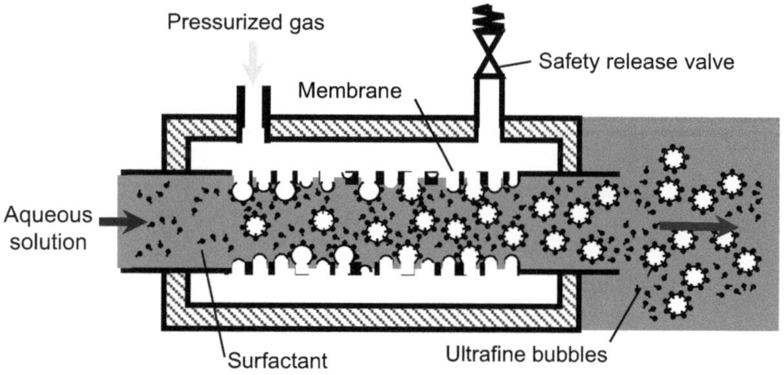

Figure 2.26 Microporous membrane type ultrafine bubble generator with adding surfactant.

Ultrafine bubbles having an average diameter of 400 nm are produced in a flowing aqueous surfactant solution at about 4 m/s by injecting gas from an SPG membrane with a minimum pore diameter of 50 nm at about 3 MPa. At this time, it is necessary to add a surfactant with a critical micelle concentration or higher to the aqueous solution to generate ultrafine bubbles. With static pure water, it is not possible to generate ultrafine bubbles with the generator. This method is excellent for applications where there is no problem even if the surfactant is contained in liquid.

2.3.4.5 Ultrasonic irradiation

When ultrasonic waves are applied to water in which a relatively poorly soluble gas such as air is dissolved, bubble generation due to gas cavitation occurs [30]. Ultrafine bubbles are generated by directly contacting an ultrasonic horn that transmits ultrasonic waves of an appropriate frequency and output with water [33] or an aqueous solution [16] containing dissolved gas, as shown in Fig. 2.27. In order to produce a large volume of ultrafine bubble water by the ultrasonic irradiation method, electric power for generating large ultrasonic waves is required. However, it is suitable for producing a small amount of ultrafine bubble water or a relatively high concentration of ultrafine bubble water.

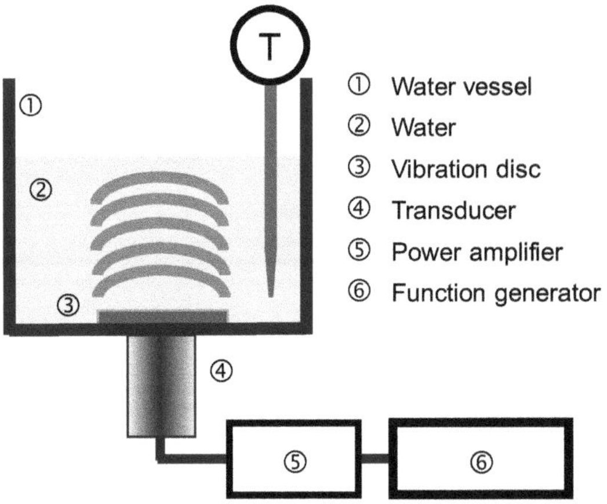

Figure 2.27 Ultrafine bubble generation with direct irradiation of ultrasound into liquid.

2.3.4.6 Strong shaking

A strong shaking method has been developed to produce ultrafine bubbles in a completely isolated system [8, 23, 35], as shown in Fig. 2.28. The gas and liquid charged in the sealed container are set in a powerful shaker such as a cell crusher [36]. When the gas and liquid are vigorously mixed under appropriate conditions, ultrafine bubbles are produced. Although this is a relatively new method, it has the feature that it can produce small quantities of ultrafine bubbles in a batch process.

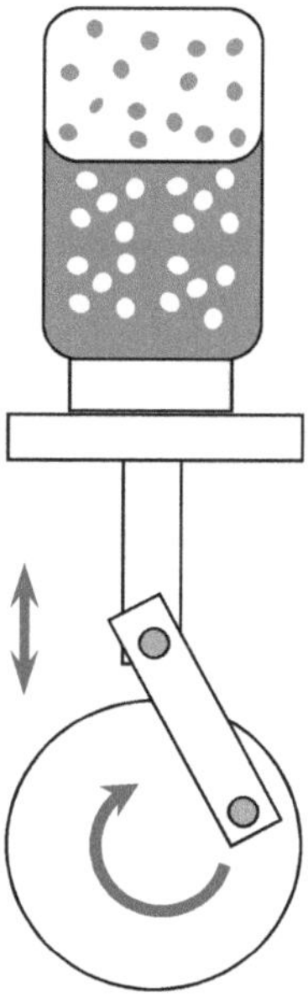

Figure 2.28 Ultrafine bubble generation with strong shaking.

2.4 Measurement of Ultrafine Bubbles

The measurement techniques for ultrafine bubbles are still limited compared to those for microbubbles. Most of the measurement techniques for ultrafine bubbles, which are now in widespread use, are derived from or improved upon the principles and instruments used to measure solid nanoparticles suspended in liquid. For each, the cost is high, and there is still room for improvement. In this section, however, we will learn them as essential tools for the development of ultrafine bubble science. The measurement techniques frequently used for measurement at present will be discussed. Because nanoparticle measurement techniques are advancing day by day, readers who are particularly interested in fine bubble measurements should be encouraged to survey the latest papers and technical reports.

2.4.1 Size and Number Concentration

The size and number concentration of ultrafine bubbles are fundamental information in evaluating the degree of various properties of ultrafine bubble water. In water, ultrafine bubbles often have a mode diameter of around 100 nm. Within the range of reported size and number concentrations, ultrafine bubble water appears only clear to the naked eye and cannot be distinguished from "ordinary water" because of their small size. However, we would like to know roughly how much ultrafine bubbles is in the sample water. For this reason, various methods have been tried, including a method of observing scattered light by irradiation with a laser beam, a method of measuring the electrical resistance of an aqueous electrolyte solution by the presence of ultrafine bubbles, and a method of observing ultrafine bubble water by an electron microscope after freezing.

However, there are currently no satisfactory size and number concentration measuring instruments. Therefore, it is necessary to select a measurement method or to complement each other with a plurality of measurement methods depending on the application and cost.

2.4.1.1 Particle tracking analysis

Particle tracking analysis (PTA) or nanoparticle tracking analysis (NTA) is one of the most common measurement methods for ultrafine bubble characterization so far. The PTA estimates the number concentration and hydrodynamic size of nano and submicron particles suspended in a liquid.

Fine particles suspended in a liquid exhibit irregular Brownian motion due to the instantaneous inhomogeneity of the collision of liquid molecules with the particles. As particle size increases, the collision of liquid molecules becomes spatially uniform, and little Brownian motion occurs for particles larger than a few microns. Ultrafine bubbles, which are gas particles of less than 1 micron in size, exhibit Brownian motion in the same way as solid particles. As described above, the ultrafine bubbles can emit scattered light at the wavelength of the laser.

Figure 2.29 shows the working principle of PTA. First, a sample cell made of glass and having a known volume is filled with a transparent liquid containing ultrafine bubbles. Next, the glass cell is irradiated with a horizontal two-dimensional (x-y) planar laser sheet. Ultrafine bubbles in the cell move randomly in three dimensions by Brownian motion. Since the ultrafine bubble intercepts the horizontal two-dimensional planar laser sheet light and scatters the laser light, the scattered light can be photographed with a video camera with a microscope, which observes the plane from directly above.

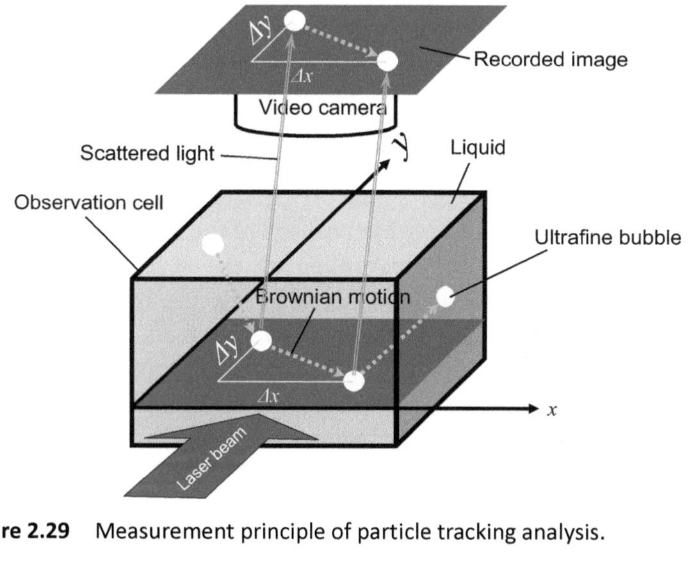

Figure 2.29 Measurement principle of particle tracking analysis.

The mean of the square displacement of ultrafine bubbles moving in the x-y plane during the unit time Δt is called the mean square displacement amount in the two-dimensional plane and is expressed by δ^2. When the ultrafine bubble displaces in the x-y plane, as shown in Fig. 2.29, δ^2 is given by the Pythagorean theorem as follows:

$$\delta^2 = \langle \Delta x^2 + \Delta y^2 \rangle \tag{2.10}$$

The symbols "⟨ ⟩" represent an average.

When the liquid bulk is at rest, the mechanism by which the solute gradually diffuses until it reaches a uniform concentration throughout the liquid bulk is also due to Brownian motion. In other words, the Brownian motion in the two-dimensional plane is connected to the diffusion coefficient D because the phenomenon in which matter moves by Brownian motion regardless of other forces (primarily gravity and forced mixing) is equivalent to diffusion.

$$D = \frac{\langle \Delta x^2 + \Delta y^2 \rangle}{4\Delta t} = \frac{\delta^2}{4\Delta t} \tag{2.11}$$

where D is the diffusion coefficient [m^2/s] and is given by the following Stokes–Einstein equation:

$$D = \frac{k_{\text{Boltz}} T}{3\pi \mu d_{\text{B}}} \tag{2.12}$$

where k_{Boltz} is the Boltzmann constant ($= 1.38064852 \times 10^{-23}$ J/K), μ is the liquid viscosity [Pa · s], T is the absolute temperature [K], and d_{B} is the ultrafine bubble diameter [m]. By transforming Eq. 2.12, the ultrafine bubble diameter d_{B} can be obtained.

In the PTA, the intensity of the scattered light (brightness) is not evaluated because the actual size of the ultrafine bubble is not directly related with it. Only the position of the light point that emits the scattered light is photographed by the video moving image, and only the effective light point that drew the locus on the two-dimensional (x-y) plane is statistically analyzed. In this method, relatively well-structured and reproducible data can be obtained by utilizing the property in which Brownian motion is active. In addition, since a simple principle is used, the theoretical reliability is high, and it has the advantage that the analysis can proceed while confirming the video motion of the Brownian motion.

However, since Eq. 2.12 does not include a variable having a large difference in value between the solid and the gas, it is not possible to discriminate between an ultrafine bubble moving in a sample solution and a solid fine particle.

On the other hand, in this method, since accurate measurement of Brownian motion is hindered when the liquid bulk is not at rest, it is not possible to directly measure or continuously measure the diameter and number concentration of ultrafine bubbles in a container such as a water tank, and a small amount of samples must be taken and applied to a nanoparticle Brownian motion tracking system in a batch mode.

2.4.1.2 Dynamic light scattering

Dynamic light scattering (DLS) has been used for decades and is the most popular size-measuring technique for nano and submicron particles. When ultrafine bubbles in a fixed volume of sample are irradiated with laser light, scattered light is emitted as described above, but the scattered light intensity fluctuates due to Brownian motion. When the diameter of ultrafine bubbles is small, the intensity of scattered light largely fluctuates with time because the velocity of Brownian motion is large. On the other hand, when the diameter of the ultrafine bubble is large, the fluctuation of the scattered light intensity becomes small because the Brownian motion is small.

Figure 2.30 shows a measurement system of the DLS method. The laser beam is irradiated to the ultrafine bubbles in the sample cell, and the scattered light whose intensity changes with time passes through the pinhole and is captured by the photomultiplier tube. And then the time fluctuation of the scattered light intensity is measured.

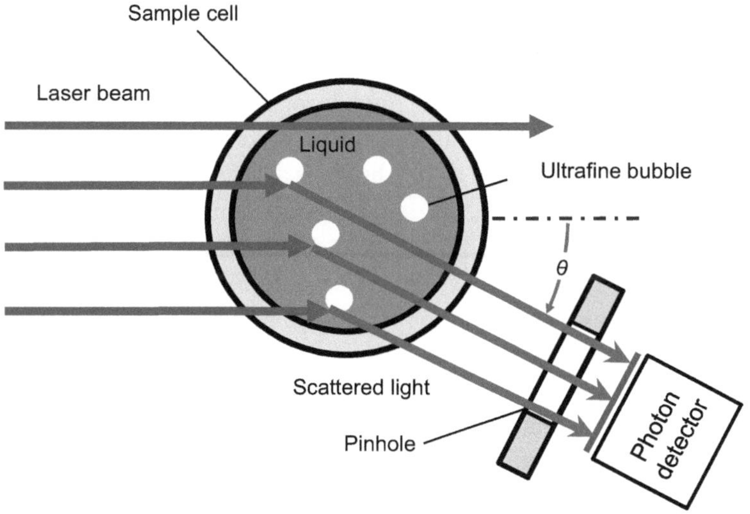

Figure 2.30 Measurement principle of dynamic light scattering.

In general, the fluctuation of the scattered light intensity is analyzed by the photon correlation method.

The first-order autocorrelation function is

$$G_1(\tau) = \exp(-Dq_0{}^2\tau) \tag{2.13}$$

The second-order autocorrelation function is

$$G_2(\tau) = 1 + \xi|G_1(\tau)|^2 \tag{2.14}$$

where τ is the time, q_0 is the scattering vector, D is the diffusion coefficient, and ξ is the device constant. Therefore, if $G_1(\tau)$ and $G_2(\tau)$ are obtained from fluctuations in the measured scattered light intensity, the diffusion coefficient D is obtained, and the ultrafine bubble diameter d_B is obtained from Eq. 2.12 by modifying the Stokes–Einstein equation.

However, as with the PTA, this technique cannot distinguish solid nanoparticles from ultrafine bubbles, and therefore, it is necessary to avoid contamination of the sample with foreign substances that cause Brownian motion other than ultrafine bubbles. Further, DLS requires relatively higher number concentration of ultrafine bubbles ($> \sim 1 \times 10^8$/mL) to obtain reproducible results. Because of this problem, reliable and reproducible measurement becomes difficult with dilute ultrafine bubble water.

2.4.1.3 Laser diffraction and scattering method

The laser diffraction and scattering method utilizes the fact that the three-dimensional scattered light intensity distribution depends on the diameter of fine bubbles. Light scattering occurs when light parallel to particles such as fine bubbles and fine solids is irradiated.

When the particle size is very large compared to the laser wavelength, the scattered light intensity is proportional to the square of the particle size. Diffraction is dominant for microbubbles larger than several 10 μm, so the microbubble diameter distribution is measured by diffraction pattern analysis based on the Fraunhofer diffraction theory.

On the other hand, in the case of microbubble and relatively large ultrafine bubble, scattering becomes dominant, and therefore, a scattering intensity pattern analysis based on the Mie scattering theory is performed. In the smaller ultrafine bubbles, Rayleigh scattering occurs, and the scattered light intensity distribution (change in scattered light intensity with respect to the observed angle) becomes not only asymmetric in the traveling direction (forward) of the laser light and the light source direction (backward) but also complicated distribution. Therefore, the measurement accuracy and resolution deteriorate, but recently, the lower limit of measurement has been extended to several 10 nm by utilizing the difference in scattering characteristics due to the shortening of wavelength and the difference in polarization direction.

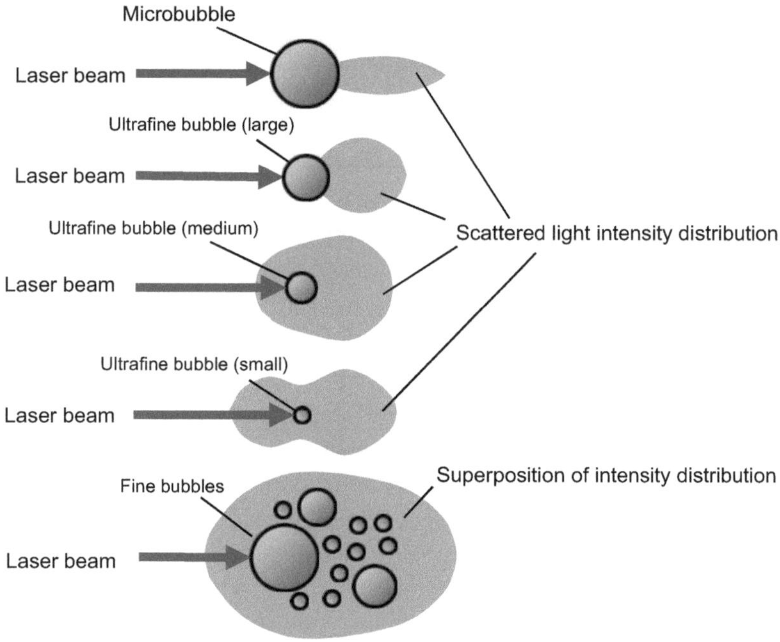

Figure 2.31 Measurement principle of laser diffraction and scattering method.

Figure 2.31 shows the principle of the laser diffraction and scattering method [12]. When a laser beam is irradiated on a fine bubble floating in a liquid, diffracted and scattered light concentrates in the traveling direction (forward) of the laser beam in a microbubble of comparatively large diameter. When an ultrafine bubble of 1 micron or less is formed, diffracted and scattered light patterns not only spread in the front but also slightly generated in the periphery (lateral). In the smaller ultrafine bubbles, the diffracted and scattered light intensity increases in the direction close to the laser light source (backward), and in the smallest ultrafine bubbles, the diffracted and scattered light also spreads forward and backward to show a gourd-like pattern. When a laser beam is irradiated on a sample mixed with fine bubbles having various diameters, superposition of each pattern occurs. When the patterns of diffracted and scattered light emitted from the fine bubbles are analyzed by a computer, the ratio of the fine bubbles having respective diameters to the whole fine bubbles is obtained.

The laser diffraction and scattering method has a wide measurable range of fine bubble diameters among various fine bubble characterization methods. However, since the size distribution of fine bubbles is calculated from the intensity distribution of scattered light from the fine bubbles, only the percentage of fine bubbles of a specific size in all fine bubbles can be determined. Therefore, it is not suitable for measuring the number concentration. However, recent techniques, such as quantitative laser diffraction (qLD) [37], have been improved by computer analysis of signal strength.

2.4.1.4 Coulter method

Figure 2.32 shows the measurement principle of the Coulter method [4] (electrical sensing band method). Since the Coulter method uses a change in the electrical resistance of a sample solution, when the fine bubble water to be measured does not contain an electrolyte, the electrolyte is added to the fine bubble water to a predetermined concentration and then poured into the sample container. A glass tube having one pore (aperture) is immersed in the sample container, and the inside of the glass tube is slightly reduced in pressure by a vacuum pump, and fine bubble water in the sample container is fed to the inside of the glass tube at a constant flow rate through the aperture. An electrode is installed between the inside of the glass tube and the outside of the glass tube to measure the difference in electrical resistance between the inside and outside of the glass tube. Electrolyte is added to the sample solution to allow current to pass through. Since the cross-sectional area of the aperture is very small, most of the electrical resistance between the electrodes is derived by the aperture. Since the gas in the fine bubble does not pass the electric current, when the fine bubble is entrained with the liquid and passes through the aperture, the electric resistance increases by the area when the fine bubble passes through the aperture. As the fine bubble passes through the aperture, the electric resistance returns to the original value.

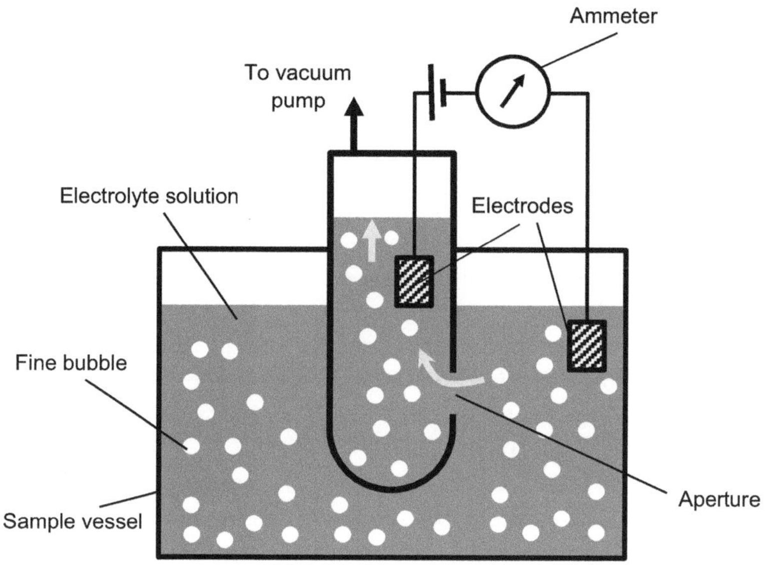

Figure 2.32 Measurement principle of Coulter method.

One fine bubble can be counted at one peak of the temporal change in electric resistance, and the diameter of the fine bubble can be calculated based on the maximum value of the resistance peak. However, generally an adjustment of experimental parameters is required so that fine bubbles pass through the aperture one by one. In order to increase the accuracy, it is also necessary to change the aperture size of the glass tube according to the range of the fine bubble diameter existing in the sample solution. It is possible to easily calculate the number concentration of fine bubbles from the number of fine bubbles passed and the flow rate of sample liquid passing through the aperture.

This method is very simple in principle and was the first successful method for measuring the diameter and number concentration of ultrafine bubbles. The reduction in aperture size enables small fine bubble to be measured. However, it is necessary to always add an electrolyte to water, and it is not possible to measure a liquid containing a fine bubble with low electrical conductivity or a fine bubble water in which an electrolyte cannot be added.

2.4.1.5 Quick-freezing replica method

As an important measuring technique capable of observing very fine particles of 1 micron or less by photographing, we can use the electron microscope photographing technique. It is very often used to observe the surface of solid particles, but it is impossible to observe liquids as well as gases. The chamber containing the electron microscope sample must be vacuumed so that electrons do not collide with gas molecules. Since the liquid evaporates under vacuum, the ultrafine bubble that exists surrounded by the liquid cannot be maintained.

Therefore, a technique was devised to make a replica in which the liquid around the ultrafine bubble is replaced with a metal thin film, and to observe the surface shape and size. This is called the quick-freezing replica method.

Figure 2.33 shows the principle and procedure of the quick-freezing replica method [9]. A sample of ultrafine bubble water is dropped onto a metal block plate cooled with liquid helium at −269°C. The water surrounding the ultrafine bubbles immediately freezes into an ice phase (solid water), which confines the ultrafine bubbles. The frozen sample is cut with a diamond knife to expose the location of the ultrafine bubbles as a recess. This surface is platinum deposited with an electron gun, and a thin film of carbon is deposited on top of it for reinforcement and then the ice phase is melted to complete the replica of the ultrafine bubble. Since the replica contains neither gas nor liquid and the surface is platinum, surface observation can be easily performed by electron microscopy, and analysis can be performed by the same method as general observation of the shape of solid fine particles and particle size distribution measurement of fine particles.

As the temperature of the liquid decreases, the solubility of the gas increases. Therefore, there is a possibility that the size or the number concentration of the ultrafine bubble, which has reached the equilibrium at a certain temperature, is lowered by the solution temperature lowering due to the progress of the dissolution. In addition, when water freezes on ice, an ice phase is formed from the contact surface with the cooling plate, and growth occurs, and ultrafine bubbles are pushed out together with unfrozen water. In order to suppress the behavior that may cause such an error, it is important to freeze the ultrafine bubble water at a normal temperature at an extremely high speed, and slow freezing treatment is inappropriate and quick freezing is essential.

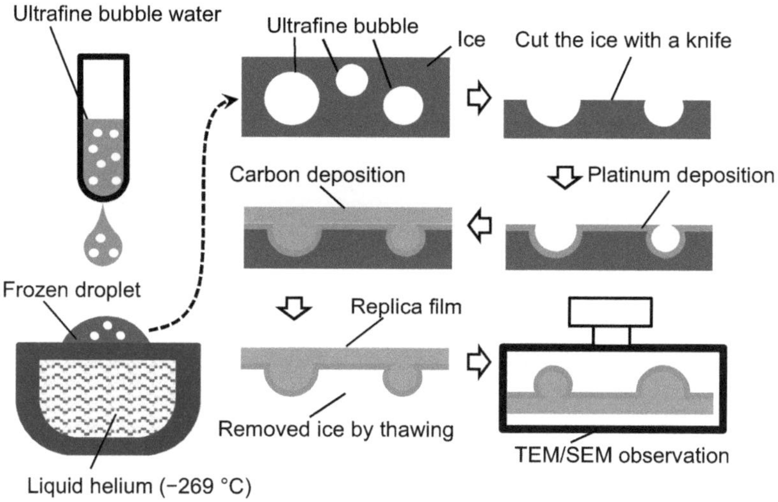

Figure 2.33 Measurement principle and procedure of quick-freezing replica method.

2.4.1.6 Resonant mass measurement

The natural frequency of an object is generally smaller when its mass is heavier and larger when its mass is lighter. Based on this physical principle, the resonant mass measurement estimates the size and number concentration of ultrafine bubbles.

The principle of the resonant mass measurement method is introduced in Fig. 2.34. A microcantilever microfabricated by the microelectromechanical systems (MEMS) technology is placed in a microchannel. The microcantilever is provided with an inlet into which a sample containing an ultrafine bubble flows at its root, a flow path in which the sample reaches the tip and turns back, and a separate outlet at its root. Steady vibration is applied to the cantilever up and down with the base of the microcantilever as a fulcrum. The natural period (or natural frequency) of the cantilever shows a relatively large difference due to a very small change in the total mass of the cantilever itself and the samples in the channel. Figure 2.34a–c schematically shows the following states:

(a) Only liquid flows into the microcantilever inner channel.
(b) Liquid containing ultrafine bubbles whose density is smaller than the liquid flows into the flow path.
(c) Liquid containing solid nanoparticles whose density is greater than the liquid enters the flow path.

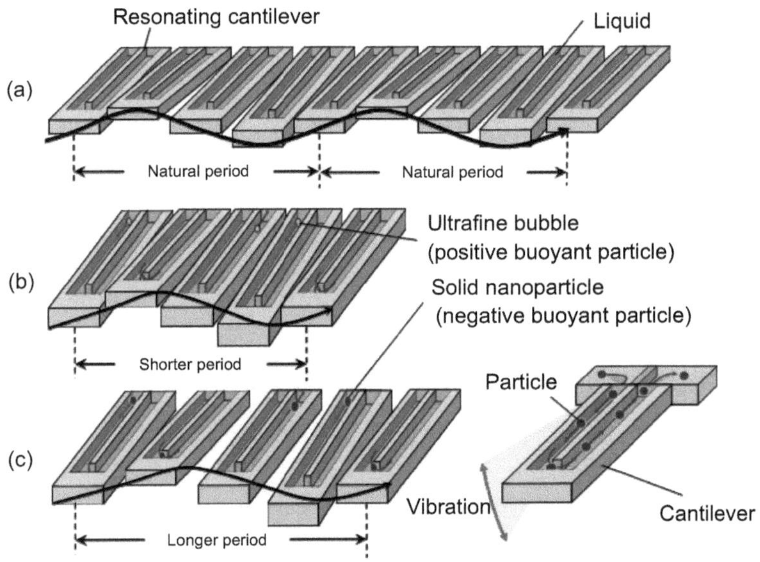

Figure 2.34 Measurement principle of resonant mass measurement.

In the above cases, since the total mass of the microcantilever itself and the sample in the channel is lighter in (b) than in (a), the natural period becomes shorter (with a large natural frequency). On the other hand, the natural period of (c) is longer, i.e., the natural frequency is smaller than that of (a) because the total mass of (c) is heavier. If the gas density in the ultrafine bubble is known, the liquid mass reduced by one ultrafine bubble can be calculated. We can also calculate the volume of one ultrafine bubble with the information of the liquid density. If the particle can be regarded as a true sphere, the diameter of the ultrafine bubble can be estimated. A single solid nanoparticle can be similarly measured. For the measurement of the distribution of ultrafine bubble diameter, each ultrafine bubble is measured and statistically processed. The number concentration of the ultrafine bubbles can be obtained by measuring the sample volume flow rate into the flow channel.

The advantage of this principle is that based on the density of the liquid (natural period or frequency), ultrafine bubbles and solid nanoparticles can be distinguished as a difference in natural period or natural frequency. On the other hand, the flow channel in the microcantilever made by MEMS is very thin, and very delicate preliminary adjustment is required such as the effect of pressurization when a sample is press-fitted, selection of the optimum ratio between the diameter of the ultrafine bubble and the width of the flow channel, sample concentration adjustment for allowing only single ultrafine bubble to flow into the microcantilever, etc. It is expected to be improved in the future.

2.4.2 Zeta Potential

It is explained that the dispersed phase surface such as fine bubbles and fine particles moves with the electric double layer and some ions around it, and the potential at the sliding surface is called zeta potential. The zeta potential is measured by electrophoresis.

Figure 2.35 shows the principle of electrophoresis for measuring zeta potential. A conductive liquid is placed in a cell where the positive and negative electrodes face each other. When an electric field is applied from both poles, positively charged particles migrate to the negative pole and negatively charged particles migrate to the positive pole, depending on their electric charges. The migration velocity u is then measured. For accurate measurement of u, methods using laser Doppler or image analysis have been put into practical use. From the value of u, the electrophoretic velocity u_{E} is given by

$$u_{\mathrm{E}} = \frac{u}{E} \tag{2.15}$$

where E is the electric field. Thus, the zeta potential ζ is determined by the following equation:

$$\zeta = \frac{4\pi\mu u_{\mathrm{E}}}{\varepsilon} \tag{2.16}$$

where μ is the viscosity of the solvent and ε is the dielectric constant of the solvent.

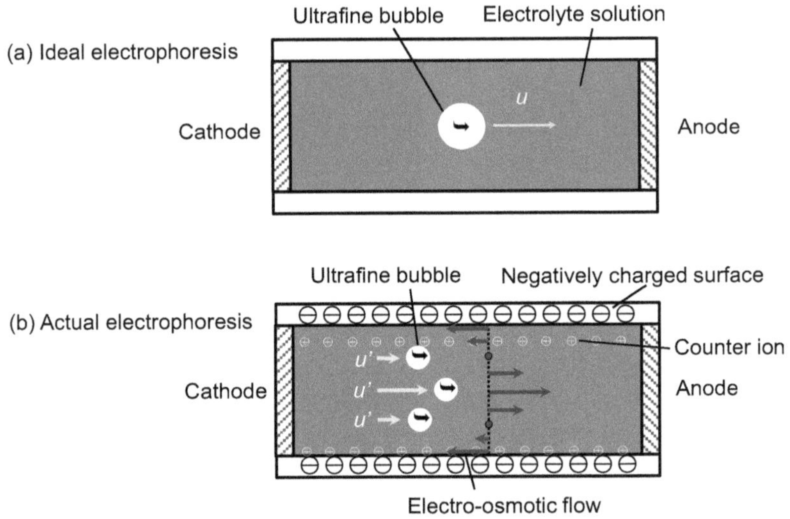

Figure 2.35 Measurement principle of zeta potential.

In the ideal case where the conducting liquid on which the particles migrate is completely stationary, the migration velocity u of the particles is equal anywhere in the cell. However, in the cell during zeta potential measurement, charging also occurs on the cell wall, so that counter ions in the solution having a polarity opposite to the polarity of charging of the wall are biased to the cell wall. When an electric field is applied to the cell, counter ions that are localized along the wall flow to the electrode side with opposite polarity. Then, in the vicinity of the center of the cell, the flow in the opposite direction is generated. This phenomenon is called electroosmotic flow.

Since the fine particles in the cell move by electrophoresis in the electroosmotic flow of the conductive liquid, the migration speed varies depending on the position from the inner wall of the cell, and these are the apparent migration speed u'. Therefore, since an equation for predicting the flow velocity distribution of the electroosmotic flow in the cell has been developed, the true electrophoretic velocity u can be obtained by correcting the apparent electrophoretic velocity u', and the zeta potential ζ can be measured.

Generally, particles for which zeta potential is measured are inorganic substances, proteins, or colloids, but measurement of ultrafine bubbles and microbubbles has been attempted. However, proper sample preparation is required for measuring the zeta potential, and when the number concentration of the ultrafine bubbles is too high or too low, adjustment such as dilution or increasing the number concentration is required. In addition, the addition of electrolyte to ultrafine bubble water may be necessary because the measurement becomes difficult in principle without the conductive liquid phase. In the measurement of the zeta potential of the microbubbles, the rising velocity of the microbubbles is large with respect to the electrophoretic velocity, so that the motion of the microbubbles cannot be captured in a small electrophoretic cell. Therefore, a measurement method has been developed in which microbubbles are raised in an electrophoretic cell having a horizontal electric field and sufficient height, and only the horizontal velocity component is separated from the motion of the microbubbles to obtain the electrophoretic velocity [24].

References

1. Aya, N. (2015). Fine bubble basics and application to cleaning, Special issue application to fine bubble cleaning, industrial cleaning, Japan Industrial Cleaning Council, No.16, pp. 8–16 (in Japanese).
2. Clift, R., Grace, J. R., and Weber, M. E. (1978). *Bubbles, Drops and Particles*, Academic Press.
3. Gasiorowski Denis, E. (2014). The fine bubble breakthrough, ISO News, available at www.iso.org/news/2014/05/Ref1844.html.
4. Graham, M. D. (2003). The coulter principle: Foundation of an industry, *Journal of the Association for Laboratory Automation*, **8**(6), pp. 72–81
5. Einstein, A. (1905). Über die von der molekularkinetischen Theorie der Wärme geforderte Bewegung von in ruhenden Flüssigkeiten suspendierten Teilchen. *Annalen der Physik* (Wiley-VCH Verlag), **322**(8), 549–560.
6. Fujiwara, A. (2006). Microbubble generation method using Venturi tube, *Monthly Eco-Industry*, **11**(3), pp. 27–30 (in Japanese).
7. Fuki Seisakusho Co., Ltd. (2019). www.fuki-ss.co.jp/wp-content/uploads/microstar.pdf (in Japanese).

8. Jin, J., Feng, Z., Yang, F., and Gu, N. (2019). Bulk nanobubbles fabricated by repeated compression of microbubbles, *Langmuir*, **35**, pp. 4238–4245.
9. Kawasaki, K., Kawasaki, T., Kobayashi, H., Maeda, S., Nishihara, I., and Fujita, T. (2015). Adhesion of nanoparticles on surfaces of ultrafine bubbles observed by quick-freezing replica electron microscopy, *Proceedings of Japanese Society for Multiphase Flow Symposium*, E114 (in Japanese).
10. Kukizaki, M., Nakajima, T., Sou, G., and Ohama, Y. (2004). Generation of monodisperse nanobubbles by porous glass membrane and bubble size control, *Kagaku Kogaku Ronbunshu*, **30**, pp. 654–660 (in Japanese).
11. Makuta, H. (2017). Microbubble generation technology using ultrasonic waves, *Journal of the Acoustical Society of Japan*, **73**(7), pp. 417–423 (in Japanese).
12. Maruyama, M. (2013). Measurement of micro-bubbles and nano-bubbles using by laser diffraction method, *Japanese Journal of Multiphase Flow*, **27**(1), pp. 53–55 (in Japanese).
13. Miyahara, T. (2005). Generation of microbubbles by liquid flow through a packed bed, *Water Treatment Technology*, **46**(5), pp. 209–215 (in Japanese).
14. Mogami, K. and Nakata, T. (2017). Mixed processing body, mixed processing method, fluid mixer, fluid mixing processing apparatus, and fish culture system, Japanese Patent No. 6176881.
15. Nakamura, N. and Hasegawa, H. (2004). Mechanism of microbubble generation, *Proceedings of the 39th Annual Meeting of the Japan Society of Mechanical Engineers*, Tohoku Branch, No. 114, pp. 28–29.
16. Nirmalkar, N., Pacek, A. W., and Barigou, M. (2019). Bulk nanobubbles from acoustically cavitated aqueous organic solvent mixtures, *Langmuir*, **35**, pp. 2188–2195.
17. Oonari, H. (2000). Swing type fine air bubble generating device, International Patent No. WO00/69550.
18. Sadatomi, M. and Kawahara, A. (2014). Fluid mixer and fluid mixing method, Japanese Patent No. 5103625.
19. Sakata, K., Yamazaki, K., and Nakajo, K. (2007). Wastewater treatment in semiconductor factories using micro-nano bubble technology, *Environmental Purification Technology*, **6**, pp. 18–22.
20. Stokes, G. G. (1951). On the effect of the internal friction of fluids on the motion of pendulums, *Transactions of the Cambridge Philosophical Society*, **9**, Part II, pp. 8–106.

21. Serizawa, A., Inui, I., Yashiro, T., and Kuwara, Z. (2003). Laminarization of micro-bubble containing milky bubbly flow in a pipe, *Proceedings of the 3rd European-Japanese Two-phase Flow Group Meeting*, di Pontignano, Italy.

22. Shakouchi, T., Tatsumoto, T., Nishio, M., Tsujimoto, K., and Ando, T. (2007). Flow and aeration characteristics of micro bubble jet flow, *Progress in Miltiphase Flow Research*, **2**, pp. 33–38 (in Japanese).

23. Tachibata, K. (2018). Bubble-retaining agent-containing liquid and method for producing bubble-containing liquid, International Patent No. JP WO2018/101251.

24. Takahashi, M. (2005). ζ Potential of microbubbles in aqueous solutions: Electrical properties of the gas–water interface, *Journal of Physical Chemistry B*, **109**, pp. 21858–21864.

25. Takahashi, M., Chiba, K., and Li, P. (2007). Free-radical generation from collapsing microbubbles in the absence of a dynamic stimulus, *Journal of Physical Chemistry B*, **111**, pp. 1343–1347.

26. Takemura, F. and Matsumoto, Y. (2003). Microbubble generator and method, Japanese Patent Laid-Open No. 2003-230824.

27. Terasaka, K. (2007). Microbubble generation method and application to industrial equipment, *Kagaku Kogaku*, **71**(3), pp. 170–173 (in Japanese).

28. Terasaka, K. (2007). Apparatus for generating bubbles or droplets in liquid and method for generating bubbles or droplets in liquid, Japanese Patent, No. 4046294.

29. Terasaka, K. (2012). Fundamentals and measurement technology of fine bubbles and their applications, *Keynote Lecture, Proceedings of International Symposium on Fine Bubble Technology*, pp. 9–18.

30. Tomita, Y., Uchikoshi, R., Inaba, T., and Kodama, T. (2010). Ultrasound induced microbubble destruction and cavitation bubble generation, *Japanese Journal of Multiphase Flow*, **24**(2), pp. 162–168.

31. Toyooka, M., Uematsu, H., and Oshima, M. (2001). Gas–liquid mixing device, Japanese Unexamined Patent Application Publication No. 2001-62269.

32. Tsuji, H. (2008). Nanobubble generator, Japanese Patent, No. 4129290 (2008).

33. Yasuda, K., Matsushima, H., and Asakura, Y. (2019). Generation and reduction of bulk nanobubbles by ultrasonic irradiation, *Chemical Engineering Science*, **195**, pp. 455–461.

34. Yasuno, M., Sugiura, S., Iwamoto, S., and Nakajima, M. (2004). Monodispersed microbubble formation using microchannel technique, *AIChE Journal*, **50**(12), pp. 3227–3233.

35. Wang, C. Q., Zhao, H., Qi, N., Qin, Y., Zhang, X., and Li, Y. (2019). Generation and stability of size-adjustable bulk nanobubbles based on periodic pressure, *Scientific Reports*, **9**, 1118, doi.org/10.1038/s41598-018-38066-5.

36. Watanabe, A., Sheng, A., Endo, H., Feril, L. B., Irie, Y., Ogawa, K., Moosavi-Nejad, S., and Tachibana, K. (2019). Echographic and physical characterization of albumin-stabilized nanobubbles, *Heliyon*, **5**, e01907.

37. Yoneda, S., Niederleitner, B., Wiggenhorn, M., Koga, H., Totoki, S., Krayukhina, E., and Uchiyama, S. (2019). Quantitative laser diffraction for quantification of protein aggregates: Comparison with resonant mass measurement, nanoparticle tracking analysis, flow imaging, and light obscuration, *Journal of Pharmaceutical Science*, **108**(1), pp. 755–762.

Chapter 3

Micro- and Ultrafine Bubbles Observed by Transmission Electron Microscopy Using Quick-Freeze Replica Technique

Kazunori Kawasaki
Biomedical Research Institute, National Institute of Advanced Industrial Science and Technology (AIST), Ikeda, Osaka 563-8577, Japan
kawasaki.k@aist.go.jp

Freeze-fracture electron microscopy has been developed for the observation of hydrous samples and utilized mostly for biological samples such as tissues and cells. The method has much potential to be applied to non-biological samples as well. This chapter introduces electron microscopic images of microbubbles (MBs) and ultrafine babbles (UFBs) dispersed in bulk water, which were obtained with the quick-freeze replica technique, a distinguished version of preparing freeze-fracture replicas. Typical protocols for freeze-fracture electron microscopy consist of (a) freezing, (b) fracturing, (c) replication, (d) cleaning and harvesting of replica membranes, followed by (e) observation with transmission electron microscopy

Ultrafine Bubbles
Edited by Koichi Terasaka, Kyuichi Yasui, Wataru Kanematsu, and Nobuhiro Aya

ISBN 978-981-4877-59-6 (Hardcover), 978-1-003-14195-2 (eBook)
www.jennystanford.com

(TEM). Each stage can be customized according to the properties of respective samples, and according to the purposes of individual researchers.

For the visualization of MBs and UFBs dispersed in water, the bubble samples were likewise processed to prepare replica membranes. With the capability of exposing and visualizing the interior of water containing samples, freeze-fracture replica has gained a unique position distinct from any other analytical method. To understand the freeze-fracture replica TEM images properly, some characteristic technical knowledge is indispensable. The mechanism of imaging the shape of minute bubbles using freeze-fracture replica, including advantages and limitations, has been described here.

3.1 TEM Observation of MBs by Quick-Freeze Replica Technique

As an example of MBs, Sonazoid (GE Healthcare), a contrast medium for diagnostic ultrasound imaging, was tested for TEM observation by the quick-freeze replica technique. Sonazoid contains perflubutane as an internal gas wrapped up with bubble shell made of egg phosphatidylserine; the resulting MBs can be stably used for about 2 h. In this experiment, the MB dispersion was prepared according to the supplier's instruction and used within an hour for the quick-freezing stage (using a Polaron E7200 "Slammer" quick freeze unit, Watford, UK).

The procedure of the quick-freeze replica technique is schematically drawn in Fig. 3.1. A drop of the dispersion about several microliters was frozen with the metal-contact method, touching the sample to the mirror-finished copper block that was pre-cooled by liquid helium (Figs. 3.1A and B), and the frozen samples were stored in liquid nitrogen until the following freeze-fracture and replication stages (Fig. 3.1C). The metal-contact method has been considered to have the highest speed of freezing samples under atmospheric pressure. It is expected that the region of samples in the vicinity of the touching face within 10 to 20 μm could be frozen, while preventing ice crystal formation that might influence the shape of MBs.

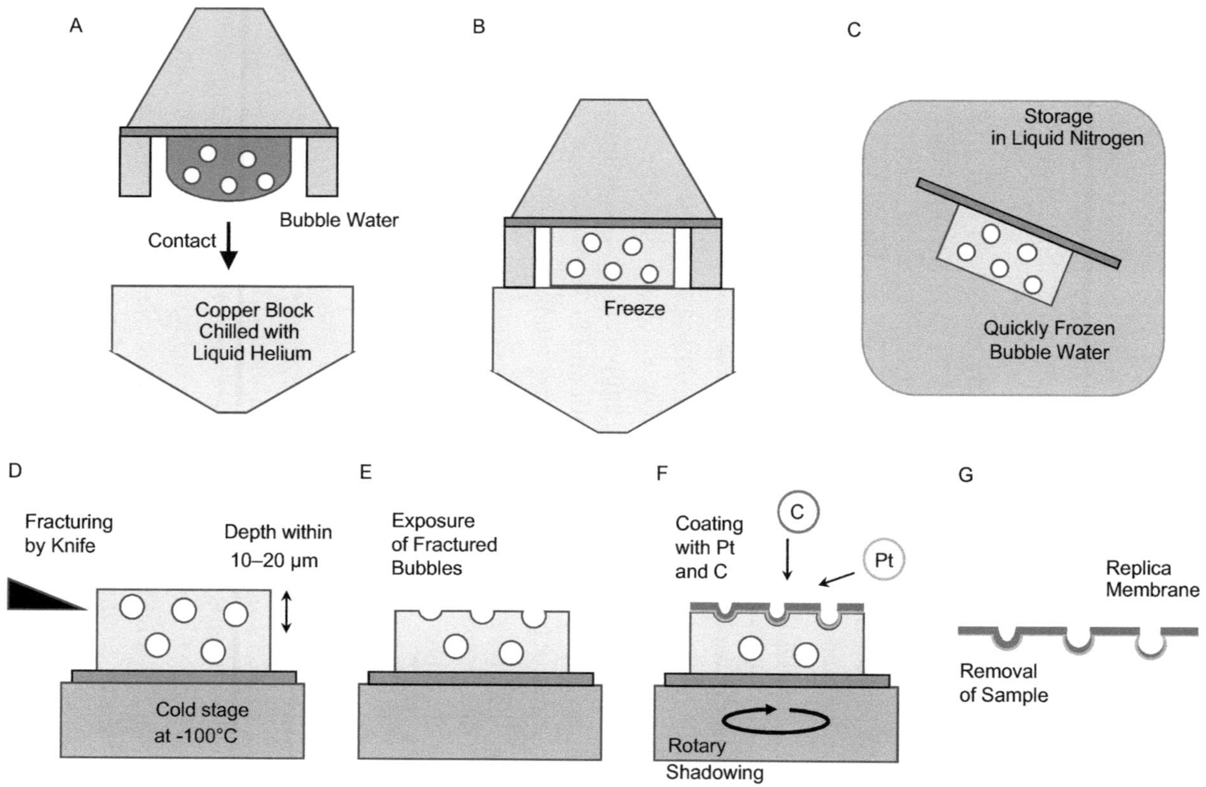

Figure 3.1 A schematic drawing of quick-freeze replica technique applied to MBs and UFBs dispersed in water. See text for details.

Freeze replicas were then prepared with a freeze-fracture apparatus (Balzers BAF400D, Liechtenstein). The quick-frozen dispersion was installed in the cold stage at −100°C in the vacuum chamber at about 1×10^{-5} Pa, so that the contact face of the frozen sample was upward (Fig. 3.1D). A knife equipped in the vacuum chamber, which was cooled with a flow of liquid nitrogen, was used to fracture the quick-frozen region of the sample at a depth of about 10 to 20 μm from the top surface (Fig. 3.1D). Fracturing was assumed to expose the lower parts of bubbles (Fig. 3.1E), and the surface was coated with platinum (Pt) about 6.5 nm thick evaporated from an electron bombardment (EB) gun situated at a low angle of 25° against the face, while rotating the cold stage horizontally for rotary shadowing (Fig. 3.1F). For reinforcement, the platinum film was further coated with carbon (C) about 25 nm thick evaporated from another EB gun at an angle of 90°, while rotating the cold stage as well (Fig. 3.1F). The sample thus processed was exposed to the atmosphere, and the Pt/C replica membrane was floated on the pure water face in a dish (Fig. 3.1G), to collect by mounting on copper grids (Veco, 150 mesh) supported with collodion films. TEM observation was performed with an FEI Tecnai G2 F20 microscope operated at an acceleration voltage 200 kV.

High-angle annular dark field scanning TEM (HAADF-STEM) mode was used for this observation. Although the choice of HAADF-STEM is not essential for the observation of freeze replicas, it has a few advantages compared to the conventional TEM mode. The most prominent feature is a Z-contrast, which is approximately proportional to the square of the atomic number of the atoms constituting the samples, Z^2. In the case of freeze replicas made of Pt and C ($Z = 78$ and 6, respectively), Pt films are much more intensively detected than C films, with a weighting factor of more than a hundred, $\sim(78/6)^2$. Since Pt films directly delineate the surface of samples and C films indirectly overlap the samples, the potential of HAADF-STEM to emphasize Pt and exclude C provides with a tremendous benefit for clearly imaging freeze-fracture replicas.

The above procedure was successfully used to obtain freeze-fracture replica images of MBs. As shown in Fig. 3.2, round shapes dispersed in water were discerned, whose diameters were typically about 2 μm, ranging from 1 μm to 4 μm. The regions of the round shapes were exclusively concaves or hollows but not convex, since the brightness of image at the regions was darker than that of the background water phase. The appearance of the round shapes in Fig. 3.2 was, thus, consistent with the two-dimensional projection images of freeze-fractured and replicated spherical caps of MBs, as schematically illustrated in Figs. 3.1E–G.

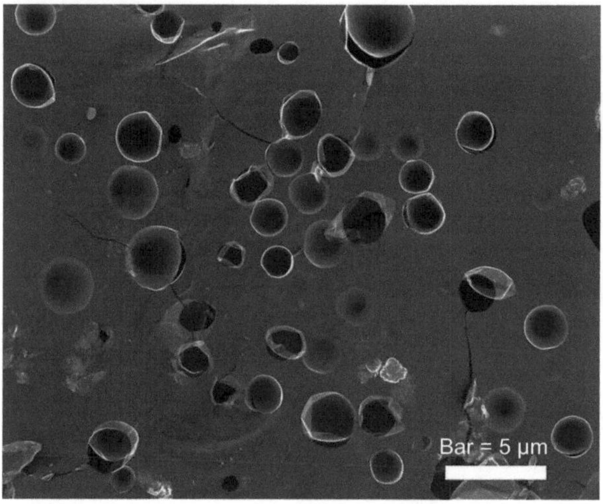

Figure 3.2 A representative HAADF-STEM image of quick-freeze replica of Sonazoid MBs. Six serial images were combined. Bar represents 5 μm.

Among the round shapes of MBs observed on replicas, some variation of appearance was noticeable, which could be divided into four categories as exemplified by HAADF-STEM images shown in Fig. 3.3. The first type of round shape, as shown in Fig. 3.3A, was like an interior of a bowl, i.e., the inner surface of a spherical cap. The second type, as shown in Fig. 3.3B, was also the inner surface of a spherical cap, which was additionally characterized with an intensively bright periphery. The third type shown in Fig. 3.3C had a torn and wrinkled film attaching to a dark round hole. The fourth type shown in Fig. 3.3D was a dark round hole without any inner face.

The four types of appearance of freeze-fractured MBs were interpreted due to the depth of fracturing level as schematically shown in Fig. 3.4. The course of freeze fracturing might encounter with various position of MBs along the depth direction (Figs. 3.4A1, B1, C1, D1), resulting in different degrees of spherical caps left behind (Fig. 3.4A2, B2, C2, D2).

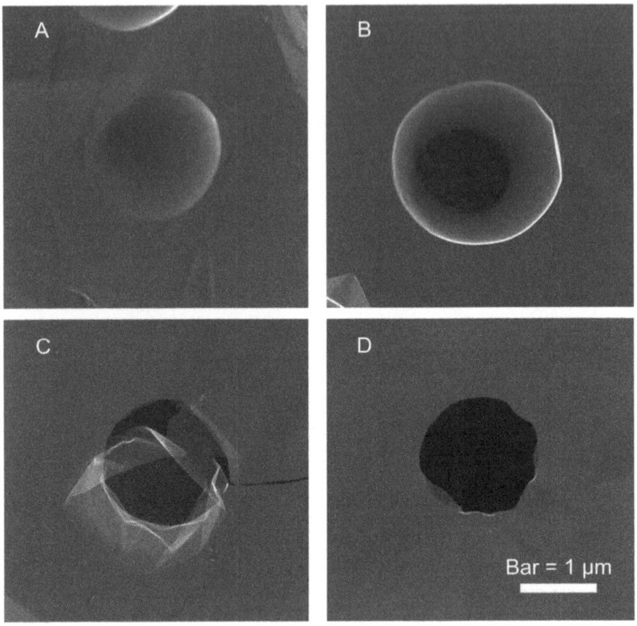

Figure 3.3 Examples of HAADF-STEM images of quick-freeze replica of Sonazoid MBs, showing variation in appearance of MBs (A–D). Bar represents 1 μm. See text for details.

At first, if fracturing exposed a spherical cap shallower than a hemisphere (Fig. 3.4A1, A2), both evaporation of Pt (from 25°) and C (from 90°) would reach the bottom of the cap, so that Pt and C films would thoroughly cover the bottom (Fig. 3.4A3). The scheme of the replica membrane in Fig. 3.4A4 is an explanation for the HAADF-STEM image in Fig. 3.3A showing an unimpaired concave face with central dark contrast. By such a shallow fracturing, the observable diameter of the round shape (spherical cap) would be shorter than that of the original MB. For example, the round shape in Fig. 3.3A appeared about 1.7 μm wide, and the diameter of the original un-fractured MB should be larger than 1.7 μm. The usefulness of images of concave faces by shallow fracturing as in Fig. 3.3A is to unveil the morphology of the inner surface of MBs, which is difficult to visualize with any other methods.

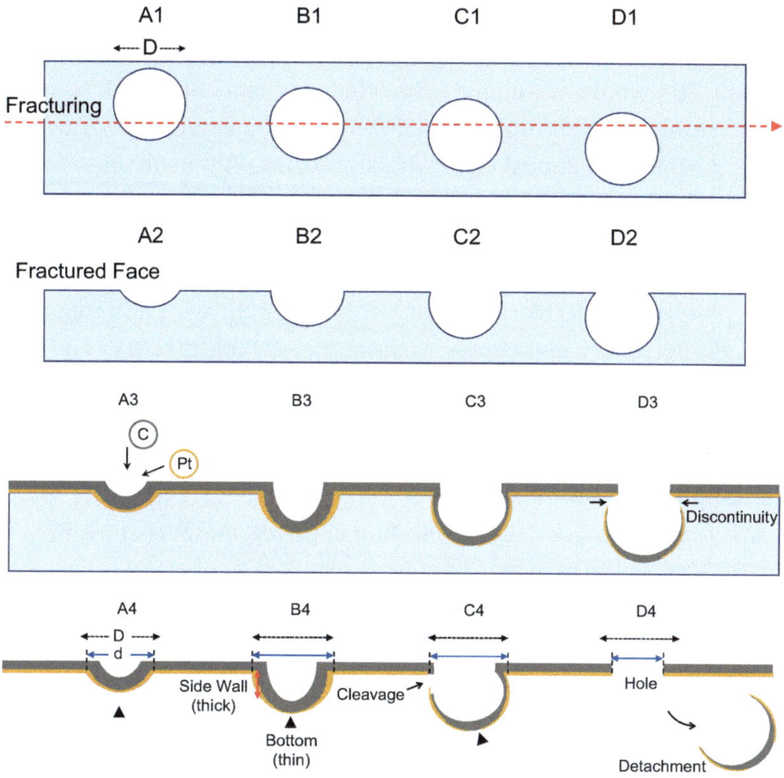

Figure 3.4 A schematic drawing of freeze fracturing and replication of quickly frozen samples of MBs and UFBs dispersed in water. (A1, B1, C1, and D1) Course of freeze fracturing encountered with various positions of bubbles along the depth direction. Long transversal arrow indicates level of fracturing. Dashed arrow marked *"D"* in A1 denotes original diameter of bubbles before fracturing. (A2, B2, C2, and D2) Variable degrees of spherical caps derived from bubbles were exposed on fractured face of frozen sample. (A3, B3, C3, and D3) Pt was coated on the fractured face by low-angle and rotary shadowing, and then C coat was superposed. Configuration of replica membranes on bubbles changed depending on the depth of exposed spherical caps. (A4, B4, C4, and D4) Removal of frozen water at room temperature completed freeze-replica specimen. Red vertical arrow in B4 denotes that side wall was projected especially thick to the direction of observation. Arrow heads in A4, B4, and C4 denote that the Pt coat at the bottom of spherical caps was thinner. Black arrows in C4 and D4 represent the cleavage of replica membrane at the most upper part of the side wall and detach of discontinuous part of replica membrane, respectively. Dashed arrow marked *"D"* in A4 and dashed arrows in B4, C4, and D4 denote the original diameter of bubbles before fracturing. Blue arrow marked *"d"* in A4 and blue arrows in B4, C4, and D4 denote the diameter of the rim of exposed spherical caps.

Second, in the case that fracturing took place almost at an equatorial face of a bubble (Fig. 3.4B1, B2), the evaporation of Pt from 25° would normally cover the side wall of bubble interior but thinly the bottom (Fig. 3.4B3). As schematically drawn in Fig. 3.4B4, by an almost equatorial fracturing, Pt film on the side wall of the fractured bubble would considerably thick along the direction of electron beam for two-dimensional transmission imaging (Fig. 3.4B4), whereas Pt film on bottom would rather thin.

The HAADF-STEM image of MB in Fig. 3.3B with an intensively bright periphery and deeply dark center was interpreted as a result of replication like the scheme in Fig. 3.4B4. It was expected that the diameter of the original MB was approximated with the diameter of round concave faces with bright periphery, as shown in Fig. 3.3B. For example, the round shape in Fig. 3.3B was about 2.2 μm wide, which was regarded a good approximation of the diameter of the MB.

Third, in the case that fracturing exposed a spherical cap slightly deeper than a hemisphere (Fig. 3.4C1, C2), the most upper part of the side wall of the bubble interior would come behind the mouth of the spherical cap. Since the angular range of evaporation from EB guns is small (straightness of evaporation), the most upper side wall was thought to be coated only faintly with Pt and C (Fig. 3.4C3). The resulting replica membrane might be easily cleaved at the mouth of the spherical cap as illustrated in Fig. 3.4C4, before or when the membrane was mounted on a copper grid. Therefore, the image of a torn and wrinkled film besides a dark hole shown in Fig. 3.3C was thought to correspond to the replica membrane once produced on MB and then deformed by subsequent operation until mounting on grids. It was inferred that a normal structure of an MB was the original state of the somewhat intricate image in Fig. 3.3C. Thus, the round dark hole with wrinkled film in Fig. 3.3C about 1.7 μm wide was attributable to an MB slightly larger.

Fourth, in the case that fracturing exposed a spherical cap corresponding to a large portion of sphere (Fig. 3.4D1, D2), concave would be so deep that both Pt and C evaporations were difficult to reach the bubble interior and the formation of Pt/C replica membrane was not ascertained. Especially, the most upper side wall of bubble interior was assumed to be masked behind the mouth of the spherical cap, resulting in discontinuity of the replica membrane between the background water phase and the side wall of bubble interior (Fig. 3.4D3). Consequently, the dark round hole without any inner face, as shown in Fig. 3.3 D, was explained as a result of so deep concave that the replica membrane at the region of MB was not stably retained, as schematically shown in Fig. 3.4D4. The round dark hole in Fig. 3.3D indicated that an MB did exist at the location. The size of the dark hole, about 1.5 μm wide, implied the original MB was considerably larger than 1.5 μm.

3.2 TEM Observation of UFBs by Quick-Freeze Replica Technique

Based on the result in MBs described above, TEM observation using the quick-freeze replica technique was expanded to UFBs. A test sample of UFBs of air dispersed in water with ultrahigh density was kindly provided by IDEC Corporation (Osaka, Japan). The density of UFBs was 8.3×10^{10} bubbles/mL, and the mean diameter was 110 nm, as determined by IDEC Corporation using a particle tracking system NanoSight NS500 (Malvern Panalytical, Malvern, UK). The reciprocal of the density of the UFB dispersion indicated that the volume occupation of a UFB was about 12 μm^3 on average. Assuming that the diameter of bubbles was uniformly 100 nm, the bubbles observable on the freeze-fracture replica membranes should have centers of those within ±50 nm from the fractured face, namely the thickness of view was about 100 nm. The volume occupation of a UFB (about 12 μm^3) divided by the thickness of view (about 0.1 μm) gave the area of occupation of a UFB (about 120 μm^2). Such estimation predicted that exploring an area of replica membranes about 11 μm × 11 μm would encounter a single UFB on average, which seemed a rewarding probability as an examination with TEM. The simple estimation on the area of occupation with a UFB highlighted the usefulness of the sufficiently high density of UFBs in the order of 10^{10}~10^{11} bubbles/mL for high-magnification studies.

The UFB dispersion was processed to prepare a quick-frozen replica and was observed by HAADF-STEM, similarly as described for MBs in the first section. Figure 3.5 shows representative images of UFBs with relatively low magnification. It was noteworthy, however, that the scale bars 500 nm in Fig. 3.5 are 10 times smaller than the bar 5 μm in Fig. 3.2, well illustrating tremendous difference in dimension between the MBs and the UFBs. From the prospects of the single UFB shown in Fig. 3.5A and three UFBs in Fig. 3.5B, there might be conveyed an impression that the bubbles were isolated or sparsely distributed. These images were, by no means, example of views with locally low density of UFBs. According to the prediction above, the incidence of detecting one or three UFBs in an area of electron microscopic frame was within expected. High-magnification images of UFBs in Fig. 3.6 showed round shapes about 110–120 nm wide. The size of UFBs observed with quick-freeze replicas was consistent with the mean diameter of 110 nm determined with NanoSight. The images suggested UFBs were essentially spherical.

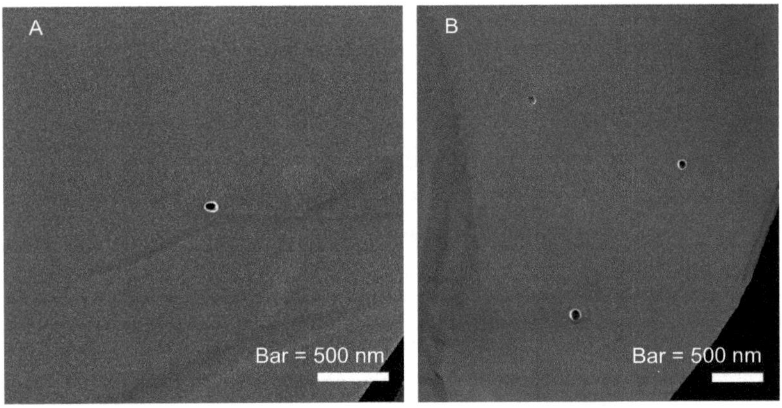

Figure 3.5 (A and B) Representative HAADF-STEM images of quick-freeze replica of UFBs dispersed in water with ultra-high density. Bar represents 500 nm.

Concerning UFBs as well, four categories of shape were found on freeze replicas as shown in Fig. 3.7. The first type in Fig. 3.7A was the inner surface of a spherical cap. The second type in Fig. 3.7B was the inner surface of a spherical cap having an intensively bright periphery. The third type in Fig. 3.7C was a dark hole with a torn and wrinkled film. The fourth type in Fig. 3.7D was a dark hole without any inner face. These four categories of appearance of UFBs in Figs. 3.7A–D were analogous to those of MBs in Figs. 3.3A–D and were interpretable due to the depth of fracturing level in the same manner as schematically shown in Fig. 3.4.

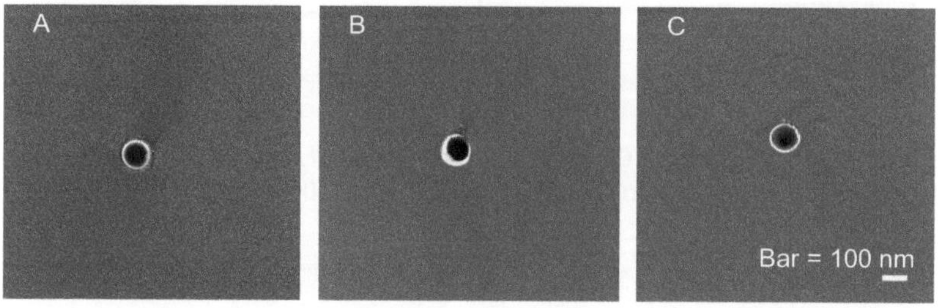

Figure 3.6 (A, B, and C) High-magnification HAADF-STEM images of quick-freeze replica of UFBs dispersed in water, showing typical UFBs with spherical shape. Bar represents 100 nm.

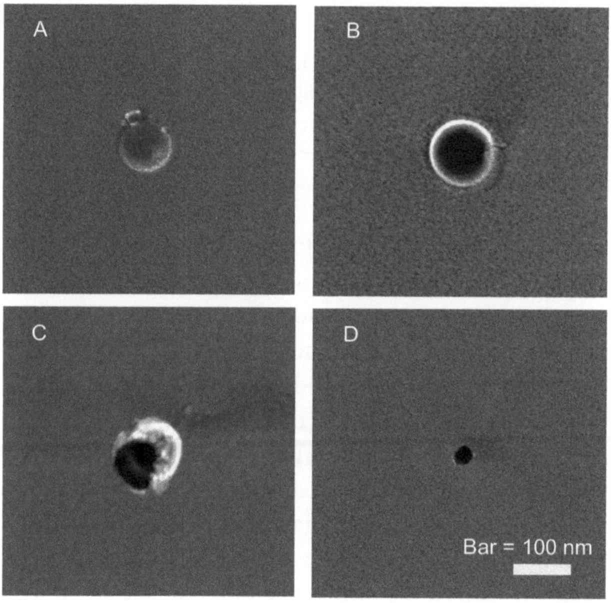

Figure 3.7 Examples of HAADF-STEM images of quick-freeze replica of UFBs dispersed in water, showing variation in round shapes of UFBs (A–D). Bar represents 100 nm.

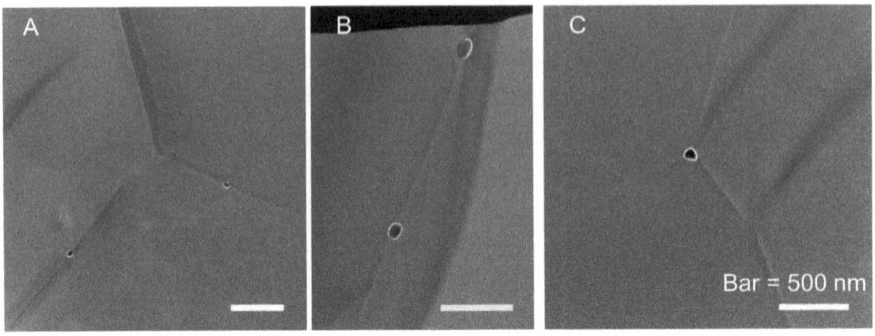

Figure 3.8 (A, B, and C) HAADF-STEM images of freeze replica of UFBs on occurrence of ice grain formation. Bar represents 500 nm.

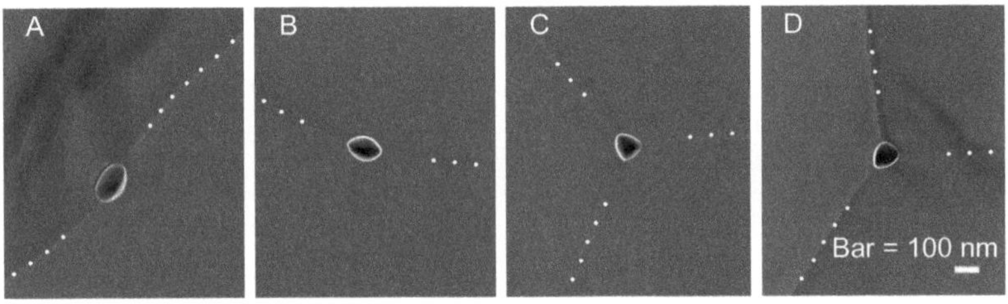

Figure 3.9 High-magnification HAADF-STEM images of freeze replica of UFBs on ice grain boundaries. (A and B) Elliptical UFBs interposed between two ice grains. (C and D) Triangular UFBs interposed at Y-shaped boundaries between three ice grains. White dots indicate position of ice grain boundaries. Bar represents 100 nm.

On the other hand, depending on the field of view of replica membranes, UFBs were found on lines as shown in Fig. 3.8. The lines represented ice grain boundaries, which could be occasionally induced in places of frozen samples even with quick-freezing experiment. The UFBs shown in Figs. 3.8A and B are located between two ice grains, but the one in Fig. 3.8C is located between three ice grains. These images suggest a tendency that UFBs were segregated from the inside of ice grains upon the formation of ice grains. Figure 3.9 shows high-magnification images of UFBs on ice grain boundaries. The UFBs between two ice grains had elliptical shape (Figs. 3.9A and B), and those between three ice grains were triangular (Figs. 3.9C and D). These images suggested that UFBs were deformed to elliptical or triangular shape presumably from the original round shape in the face of pressure by ice grains.

Hence, elliptical or triangular UFBs on ice grain boundaries might not preserve the authentic shape of UFBs, but those indicated the presence of UFBs in water dispersion. On one hand, UFBs exhibited stability against segregation from ice grains, as shown in Fig. 3.8, and on the other hand, deformability, as shown in Fig. 3.9.

3.3 Conclusion

The visualization of MBs and UFBs was possible by using quick-freeze replica electron microscopy (Figs. 3.2 and 3.5). Although the bubbles composed of gas and water are intrinsically not electron dense and unsuitable for TEM observation, by virtue of electron density of Pt replica membranes, high-contrast images of bubbles have been produced. In addition, the usage of HAADF-STEM mode, instead of conventional TEM mode, was effective to emphasize the image contrast of bubbles replicated with Pt membranes. It was ascertained that the shape of both MBs and UFBs was essentially spherical (Figs. 3.2 and 3.6). MBs and UFBs were fractured at various positions along the depth direction depending on the course of freeze fracturing, and several aspects of bubbles were revealed. The morphology of the inner surface of bubbles was visualized by shallow fracturing (Figs. 3.3A and 3.7A). Bubble periphery was observed with particularly high contrast by an almost equatorial fracturing (Figs. 3.3B and 3.7B), from which the diameter of bubbles was approximated. Round dark holes were discerned when fracturing was deeper than equatorial (Figs. 3.3C, 3.3D, 3.7C, and 3.7D). In these images, parts of replica membranes corresponding to bubbles were removed, though the holes as remaining traces illustrated the existence of bubbles dispersed in water. On occasions when the formation of ice grains was apparent, UFBs tended to be segregated from ice grains to ice grain boundaries (Fig. 3.8) and the shapes of those were deformed from spherical to elliptical or triangular (Fig. 3.9).

Acknowledgment

The author is grateful to Hideaki Kobayashi, Yuuki Kamijo, and Kazunari Araki of IDEC Corporation (Osaka, Japan) for kindly providing the valuable test sample of UFB dispersion with ultrahigh density. This work was partly supported by the Ministry of Economy, Trade and Industry (METI), Japan as part of the international standardization project for fine bubbles, and by the Council for Science, Technology and Innovation (CSTI) Japan as part of the Cross-ministerial Strategic Innovation Promotion Program's "Technologies for creating next-generation agriculture, forestry and fisheries."

References

1. Severs, N. J. 2007. Freeze-fracture electron microscopy. *Nat. Protoc.*, **2**:5473–5576.
2. Heuser, J. 1981. Preparing biological samples for stereo-microscopy by the quick-freeze, deep-etch, rotary-replication technique. *Methods Cell Biol.*, **22**:97–122.
3. Heuser, J. E. 2011. The origin and evolution of freeze-etch electron microscopy. *J. Electron Microsc.*, **60**(Suppl. 1):S3–S29.

Chapter 4

Real UFB Sample Measurements: A Few Cases

Akinari Sonoda
Health Research Institute, National Institute of Advanced Industrial Science and Technology (AIST), 2217-14 Hayashi-cho, Takamatsu, Kagawa 761-0395, Japan
a.sonoda@aist.go.jp

A few cases of real ultrafine bubble (UFB) sample measurements are introduced in this chapter, and their results are compared. First, the performance of a UFB monitor (UBM) is evaluated using reference materials and water containing UFB generated by various generators. Then, the UFB samples obtained from three different suppliers are investigated using the nanoparticle analyzer qNano to evaluate the influence of the transportation process and deterioration with time. Finally, the UBM is applied to ozone UFB water in a human safety test.

Ultrafine Bubbles
Edited by Koichi Terasaka, Kyuichi Yasui, Wataru Kanematsu, and Nobuhiro Aya

ISBN 978-981-4877-59-6 (Hardcover), 978-1-003-14195-2 (eBook)
www.jennystanford.com

4.1 Introduction

In this chapter, a few cases of real UFB sample measurements are introduced, and their results are compared. Recently, various measuring methods and devices have been developed. The laser pointer method, which visualizes the laser path based on the Tyndall effect, is the simplest method available to detect UFBs, even if the effect becomes more pronounced in contamination particles of size larger than 0.45 μm compared with the UFBs, whose mean diameter is approximately 100 nm in most cases; however, this method provides no quantitative information. Another simple instrument is the UBM, which digitizes the intensity of the scattering light passing through a cell containing a liquid sample. This instrument does not have any complex function such as data treatment or calculation; in addition, its portability is advantageous in being usable along with a UFB generator. Moreover, some sophisticated instruments are available to evaluate the properties of UFB. The nanoparticle analyzer qNano (IZON Science Ltd, New Zealand), which is based on the tunable resistive pulse sensing (TRPS) principle [1, 2], is used to measure the size and concentration of UFBs. Another instrument for fine particles equipped with an array of sensors to detect diffracted and scattered laser light has been specialized for UFBs and low concentration nanoparticles. The instrument SALD-7500nano (Shimadzu Corporation, Japan) achieves excellent sensitivity in the nanoregion, which is about 10 times higher compared with that of conventional instruments. In addition, the volume concentration (μL/L) can be converted to number concentration (particles/mL) [3, 4]; however, it is not suitable for a mixture of unknown particles, which are frequently observed in real UFB samples [5–7], because the refractive index is very important in the calculation of concentration [8]. The apparatuses employing the nanoparticle tracking analysis method, such as NanoSight (Malvern Panalytical Ltd., UK) and ZetaView (MicrotracBEL, Japan), are widely adopted for UFB measurement.

First, the performance of UBM is evaluated using reference materials and water containing UFBs is generated by different types of generators. Then, the UFB samples obtained from three different suppliers are investigated using qNano to evaluate the influence of the transportation process and deterioration with time, along with some samples from a plant factory (lettuce). The results are then compared with those of SALD-7500nano and NanoSight for reference. Finally, the UBM is applied to ozone UFB water in a human safety test.

4.2 Ultrafine Bubble Monitor

The UBM is easy to use in monitoring fine bubbles. It consists of a simple combination of flow cell, laser light, and detector without any data-handling system. In the following sections, the results obtained for polystyrene (PS) latex aqueous dispersion and water containing UFBs (UFB water) are discussed.

4.2.1 Polystyrene Standard Particle

Polystyrene latex (0.10 μm, 2.5 wt% dispersion in water) reference material was used as the sample. The latex dispersion was diluted with ultrapure water (UPW). The number concentration was estimated to be 6.0E+12 particles/mL from the measurement results of qNano. The maximum reading of UBM was 4095, and the lowest was approximately 830; however, the reading sometimes reached approximately 1000, even for only UPW, which may have been caused by some bright spots in view of the flow cell (Table 4.1).

The number concentration and the UBM readings are plotted in the linear graph shown in Fig. 4.1. The open circles and the dotted line represent the PS samples and the corresponding regression curve, respectively. The error bars were sufficiently small to be present within the diameter of the circle, except for the largest UBM reading of 3704.

In Fig. 4.2, error bars are not shown, and the log–log plot of data in Fig. 4.1 is shown. The linear relationship between number concentration and UBM reading is more clearly exhibited. The range of UBM readings (830 to 4095) corresponding to the number concentration (5E+07 to 3E+08 particle/mL for 0.1 μm PS) is not enough to measure the real UFB samples, because the range is expected to be wider.

Table 4.1 UBM reading and view of scattering light in flow cell

Sample	Value of UBM apparatus	View of flow cell
Blank ultrapure water		
150,000 dilution 4.0E+07 particles/mL	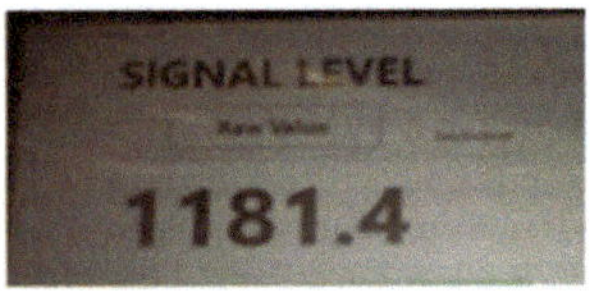	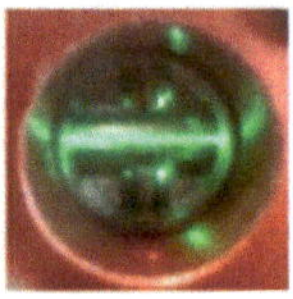
100,000 dilution 6.0E+07 particles/mL		
75,000 dilution 8.0E+07 particles/mL		
50,000 dilution 1.2E+08 particles/mL		
25,000 dilution 2.4E+08 particles/mL		

Polystyrene latex (0.10 μm, 2.5 wt% dispersion in water) dilution.

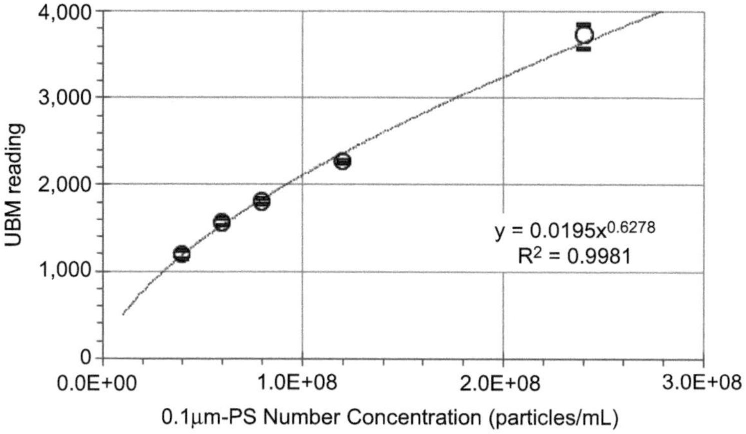

Figure 4.1 Graph showing the linear curve of PS number concentration against UBM reading.

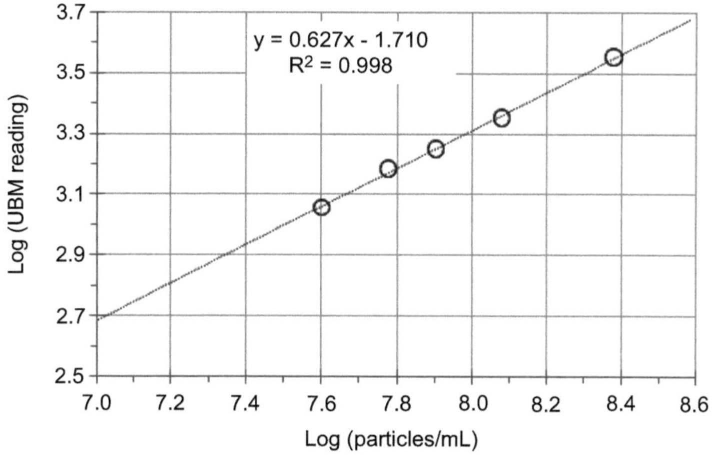

Figure 4.2 Log–Log plot of PS number concentration and UBM reading.

4.2.2 UFB Water Preparation by Agitational Mixing Type Generator

UFB water (UPW treated by agitational mixing type UFB generator) was infused into a UBM apparatus using a syringe. The reading of

UBM was in the range 1100–1200, which is slightly larger than that of UPW (blank value). On the other hand, during continuous infusion using a peristaltic pump, the UBM did not exhibit any change from the blank value; thus, it is inferred that the UFB disappears during the infusing process when a pump is used.

4.2.3 UFB Water with Contamination

We further attempted to evaluate another UFB water prepared by another type of generator using UBM. Solid particles were, however, visible to the naked eye from the sidewall of the generator tank; thus, the water was probably contaminated. The particles were recognized by the change in the color of the membrane filter (MF) (0.45 μm pore size) to gray by only 10 mL filtration of the UFB water. As demonstrated in Fig. 4.3, the color density continuously enriched each time the entire water in the tank circulated throughout the generator (which is defined as a pass). Thus, it can be inferred that the particles originated from the generator. Water containing UFB generated by 90 passes was evaluated by the UBM following filtration using membrane filters of different pore sizes. As illustrated in Table 4.2, the UBM reading decreased with the decrease in pore size, and the sample filtered using the finest filter of pore size 0.2 μm provided a reading lower than that of UPW exhibited in Fig. 4.1, implying that the bubbles were removed along with the particles during filtration. These results indicate that the contamination caused by a UFB generator significantly influences the measurement of UFB.

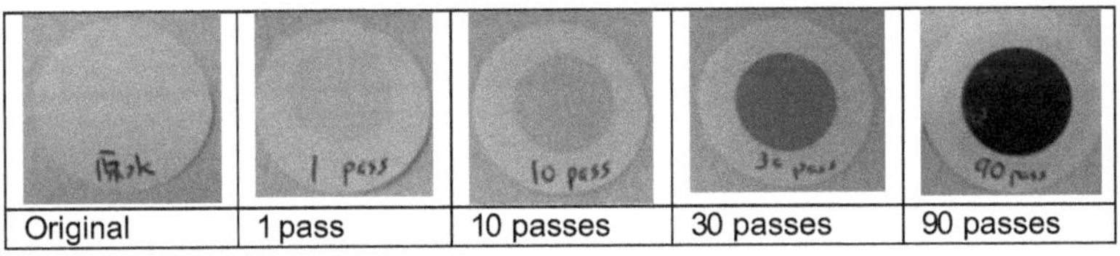

Figure 4.3 Increase in contamination caused by the UFB generator. 200 mL of UFB water was filtered by a 0.45 μm MF (ϕ: 25 mm).

Table 4.2 Results of UBM measurements for the filtered UFB water (90 passes)

Sample	Value of UBM apparatus	View in flow cell
0.8 μm IP*-MF treatment	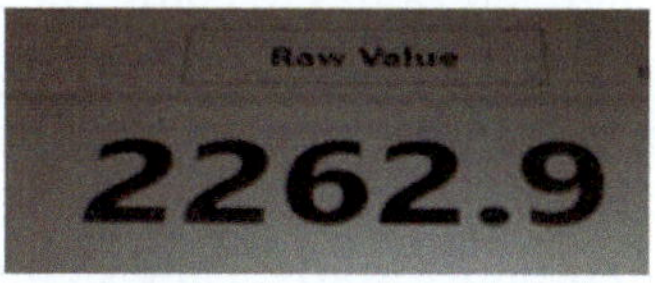	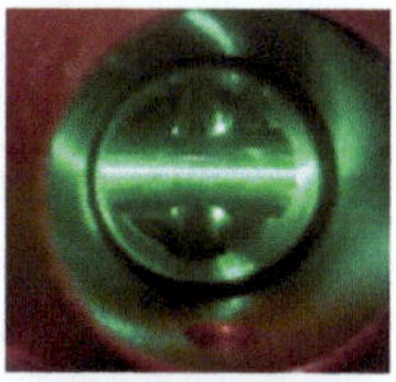
0.45 μm MF treatment	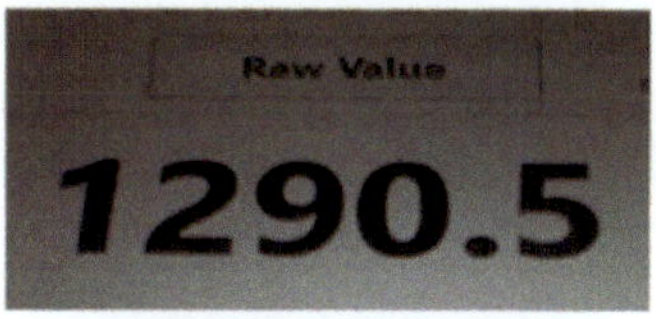	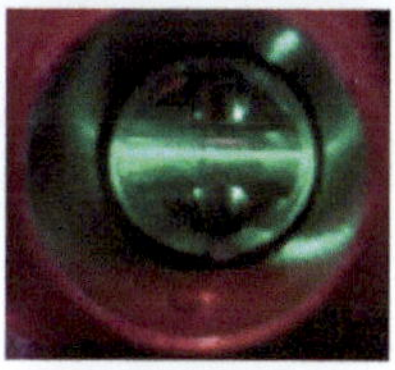
0.2 μm IP*-MF treatment		
Blank	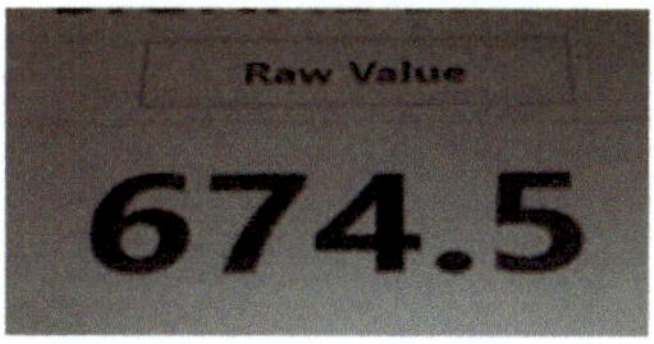	

*IP indicates iso-pores composed of polycarbonate film.

4.3 Measurement of UFB Using qNano

qNano is one of the particle measurement systems that use the TRPS principle. In this section, some measurement results of UFB using qNano were compared with those of NanoSight, which is one of the most widely used instruments employing the nanoparticle tracking analysis method. Because the qNano requires a sample containing electrolytes, a small amount of highly concentrated KCl solution was added up to 0.1 mol/L. Nanopores (NP) were installed in the qNano system to count the particles; the NP is a flexible polyurethane membrane with conical through-hole. The three-digit numbers in a model number (such as NP100 and NP200) indicate the suitable model for the particles of diameter 100 nm and 200 nm, respectively. They contain recommended values to count the number of particles. For example, the number of 1E+10 particles/mL was recommended for NP100, which is suitable for particles of diameter ranging from 70 nm to 100 nm. However, it must be noted that a smaller counting number is unlikely to produce statistically significant measurement results. The influence of transportation on the properties of UFB is investigated in the following section, for samples obtained from three different suppliers. Further, water obtained from the irrigation system of a lettuce factory is evaluated as an example of agricultural application.

4.3.1 UFB Water Provided by Keio University

UFB water produced by the high concentration bubble generator at the Keio University (Yokohama, KANAGAWA prefecture, Japan) was brought as baggage to AIST Shikoku (Takamatsu, KAGAWA prefecture, Japan) by train (mainly Shinkansen). Then the fresh UFB water was evaluated by NanoSight and SALD7500nano at the Keio University. The number concentrations measured by NanoSight and SALD7500nano were 3.5E+08 particles/mL and 1.5E+09 particles/mL, respectively. This discrepancy can be attributed to the difference in their measuring principles.

In addition, the UFB water brought from the Keio University was measured by qNano (Tables 4A.1–4A.4, Fig. 4.4) at AIST Shikoku. The tables list the results of five successive measurements. Because of a technical issue in qNano measurement, constancy of flow rate, and stability of flow, the maximum value of each measuring batch (highlighted in gray in the tables) was chosen as the representative value rather than the average. The same procedure was followed in the measurement of the three remaining samples of different elapsed days since bubble generation. The values of NP200 (Table 4A.1, 2 days, max. 86.8 particles/min) and NP100 (Table 4A.2, 9 days, max. 71.1 particles/min) were larger than those of NP200 (Table 4A.3, after 22 days, max. 2.6 particles/min) and NP100 (Table 4A.4, 14 days, max. 14.1 particles/min). The counted particles per minute were converted to particle concentration using the calibration curve of the standard sample. These values correspond to a wide range of concentration from 3.8E+06 to 6.7E+08 particles/mL. The decrease in concentration after several days implies the presence of UFB rather than solid contaminants.

Keio UFB, NP200, 2 days after measurement

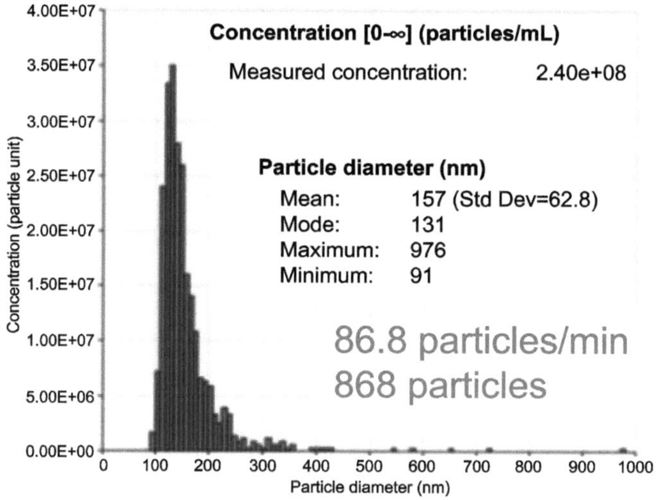

Keio UFB, NP100, 9 days after measurement

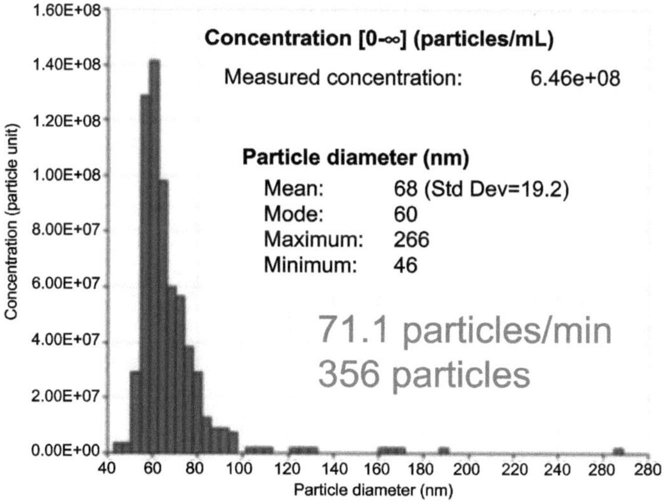

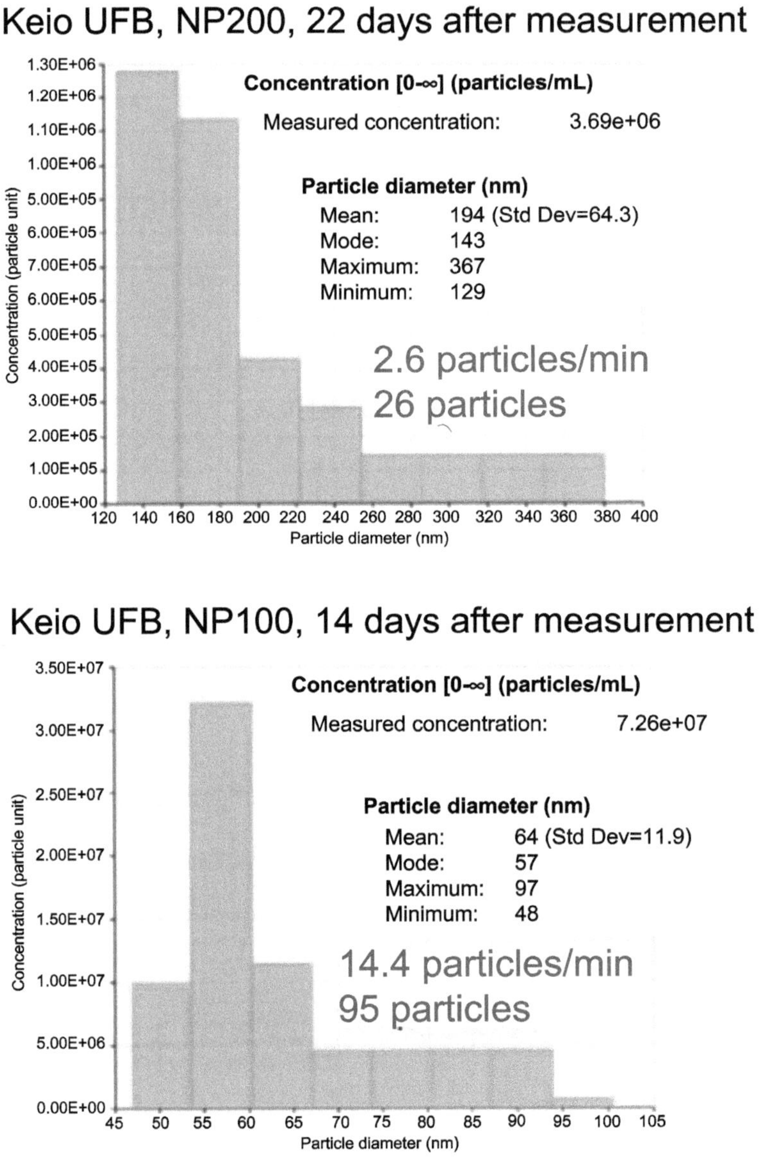

Figure 4.4 Results obtained by the qNano system for UFB water of the Keio University.

4.3.2 UFB Water Provided by Osaka University

UFB water produced by high concentration bubble generator at the Osaka University (Suita, OSAKA prefecture, Japan) was transported by delivery truck service in a plastic container (1 L) and then brought by train in glass bottles (30 mL) to AIST Shikoku (Takamatsu, KAGAWA prefecture, Japan). The UFB water was measured by the qNano system (Tables 4B.1–4B.2, Fig. 4.5).

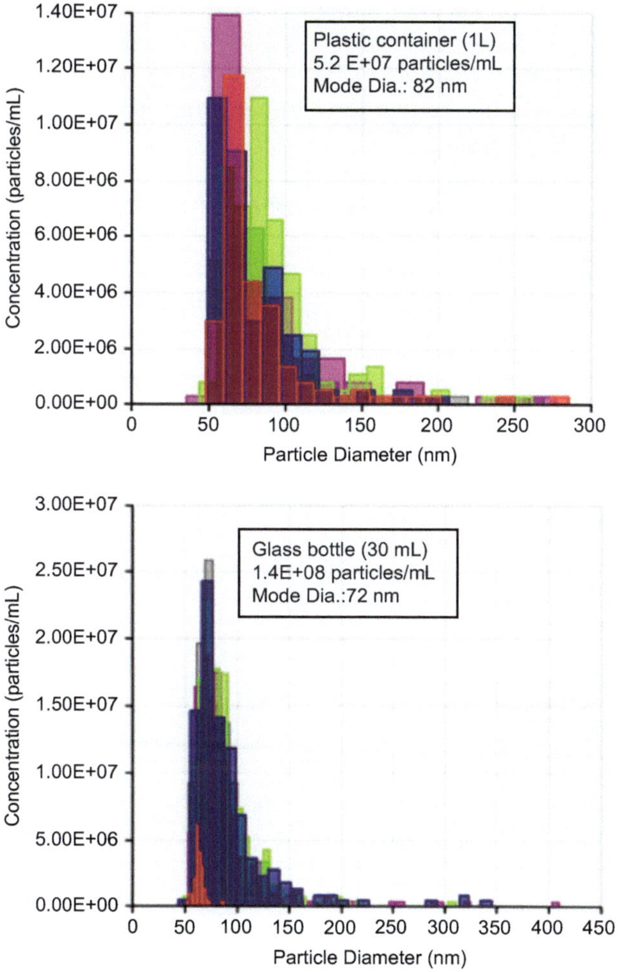

Figure 4.5 Results obtained by the qNano system for UFB water of Osaka University.

In the case of glass bottle, the concentration was about 1E+08 particles/mL, which is approximately two times larger than that of plastic container (5E+07 particles/mL).

4.3.3 Commercially Available UFB Water

Commercially available UFB water, which is prepared by high concentration UFB generator, was delivered by the supplier (Kitakyushu, FUKUOKA prefecture, Japan) to AIST Shikoku (Takamatsu, KAGAWA prefecture, Japan) in plastic bottles (1 L). A representative example of NanoSight data was attached to each bottle, which indicated that the number concentration is higher than 1E+10 particles/mL.

Table 4.3 Concentration of commercially available UFB water measured by NanoSight

	Day of generation (diluted by 50 times)		After one month (same lot unopened)		
	Concentration $/10^8$/mL	Mode Dia/nm	Concentration $/10^8$/mL	Mode Dia/nm	Remaining rate
Ultra-high concentration O_2 UFB water	156.5	59.9	6.65	91.7	4%
High concentration O_2 UFB water	80.5	83.9	2.40	59.3	3%
High concentration air UFB water	82.5	74.2	8.26	56.3	10%
High concentration N_2 UFB water	77.0	52.1	10.8	72.8	14%
Fresh ultra-high concentration O_2 UFB water	–	–	47.0	58.8	–

Nevertheless, the qNano system provided a low count of less than 10 particles/min, corresponding to a concentration less than 1E+7 particles/mL for bottled water. Further, the unopened bottles were returned to the UFB water maker and were evaluated by NanoSight (Table 4.3). The number concentration of UFB of three different gases after a month decreased in the following order: O_2 > air > N_2. Transportation and lapse of time caused reduction in the UFB number concentration, which indicates that the UFB is in a metastable state and not in a stable state.

4.3.4 UFB Water to Irrigate Lettuce at Plant Factory

The UFB water generated on site was used in a lettuce factory, where real UFB culture solutions were poured at upstream, middle stream, and downstream parts in a flow culture system. These samples were collected in glass bottles and sent through a delivery truck service. They were filtered by a syringe membrane filter unit (0.22 μm) before evaluation by the qNano system.

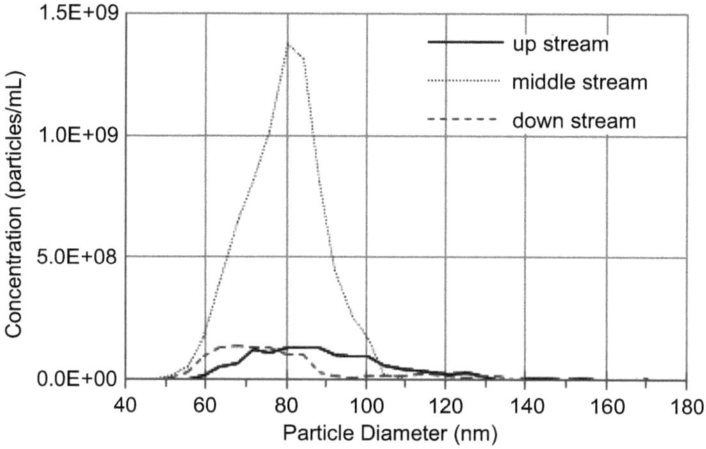

Figure 4.6 Results obtained by the qNano system for the UFB water used to irrigate lettuce.

The results obtained for NP100 (range: 70–200 nm, recommended concentration of 1E+10 particles/mL) are summarized in Table 4C.1 and Fig. 4.6. The maximum value of the upstream sample was 1.3E+09 particles/mL, the value of the middle stream increased to 7.5E+09 particles/mL, and that of the downstream decreased to 1.1E+09 particles/mL. This decrease may have been caused by the blocking of larger particles by the Nanopores. The C4 and C5 samples were counted only 1 and 2 particles, respectively, during measurement time (10 min) (Table 4C.1). Further, an increase and a decrease in the number concentration were observed along the flow path of the irrigating system. These results, however, did not result in a higher concentration of UFB, because the flow culture system is probably a contamination source with sources such as solid substances originating from the inner surface of the irrigation tube.

4.4 Human Safety Test of Ozone UFB Water

The ozone UFB water was evaluated for human safety using 3D tissue culture models. From the perspective of animal protection, the reconstructed human tissue culture models, such as skin regeneration models, are extremely useful in evaluating the safety and efficacy of poorly soluble materials and difficult-to-test formulations; the requirement of such models has increased in recent years. Moreover, the reconstructed human skin models and corneal epithelium models can be used as alternatives to animal tests to evaluate the irritants. The test was conducted at two different testing laboratories as mentioned below to verify the reliability of the testing method.

4.4.1 Preparation of Ozone UFB Water

The testing apparatus and the conditions are listed below, and a few are shown in Figs. 4.7 and 4.8.

- O_3 generator: ED-OG-RC12GC (EcoDesign Inc., Japan), O_3 generation rate: 12 g/h.
- Portable O_3 gas monitor: PG-620HA (EBARA JITSUGYO CO. LTD., Japan)
- In-line O_3 digester: ED-MD9-1500s (EcoDesign Inc., Japan)
- Flow meter: Max. 2 L/min
- Vacuum pump (for O_3 digester): 30 L/min

- Glass Water tank: Silent fit 500 (GEX. CO. LTD, Japan), W: 50 cm, D: 24 cm, H: 29 cm, 30 L
- Bubble generator: Microstar (for 100 V), FS101-S1 (Fuki CO. LTD, Japan)
- Water: purified water, model number: MGW-1 (AS ONE Corporation. Japan), TOC<0.5 mg/L, EC<2.1 μS/cm
- Teflon hose: O.D. 6 mm
- Oxygen gas cylinder with regulator
- Ultrafine bubble monitoring apparatus: UBM1.0 (Quantum Design Japan)
- Water quality meter: 2-ch Portable multi meter HQ40d (Hach) with EC&ORP probe

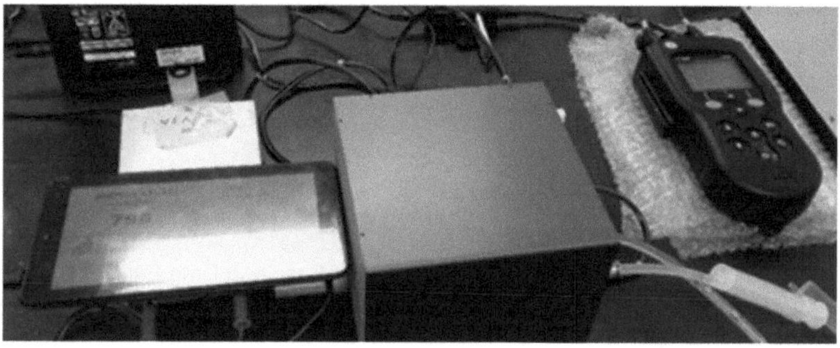

Figure 4.7 Images of the UMB1.0 apparatus and the portable multimeter.

The operating conditions of the UFB generator were as follows:

- Water level of the tank: standard value for the generator (12 mm from the lower flange, central size of generated bubble: 2 μm)
- Agitational speed of generator: 2800 rpm (50 Hz)
- Flow rate of O_3 gas: 0.5 L/min
- Pressure of O_2 gas cylinder: 0.1 MPa
- Concentration of O_3 gas: 180 g/m^3, setting controlled by auto mode

The typical analytical data of water were as follows:

- Before generation—EC: 2.44 μS/cm, ORP: 688.6 mV, UBM reading: 1386

- After generation—EC: 17.79 μS/cm, ORP: 1154.9 mV, UBM reading: 1780

4.4.2 In vitro Skin Irritation Test

The ozone UFB generator was brought to each test bench in ACEL Inc. and Food and Drag Safety Center, as shown in Fig. 4.8. It operated for about two hours to reach the saturated ozone density. Further, fresh ozone bubble containing water was used for the skin irritation test. In Fig. 4.9, the in vitro skin irritation testing setup at ACEL Inc. is shown. The tests complied with OECD TG439 (reconstructed human tissue culture models) other than the manufactures of the 3D tissue model, exposure time, and negative control (NC).

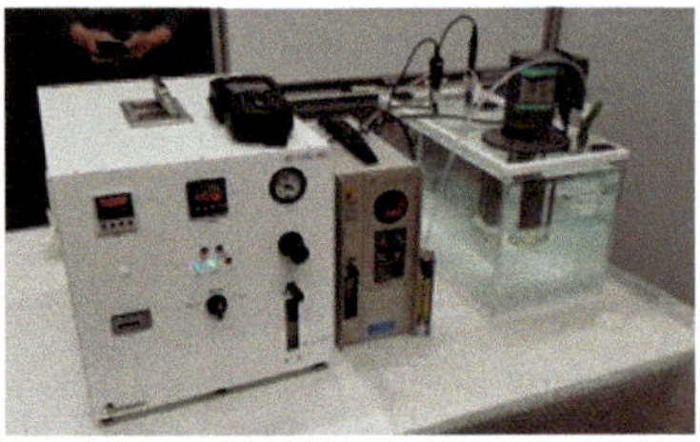

At ACEL, Inc

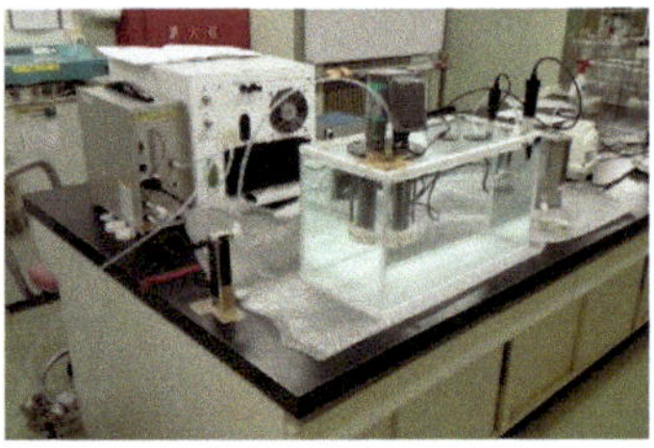

At Food and Drug Safety Center

Figure 4.8 Overview of the ozone UFB preparation setup on safety test benches.

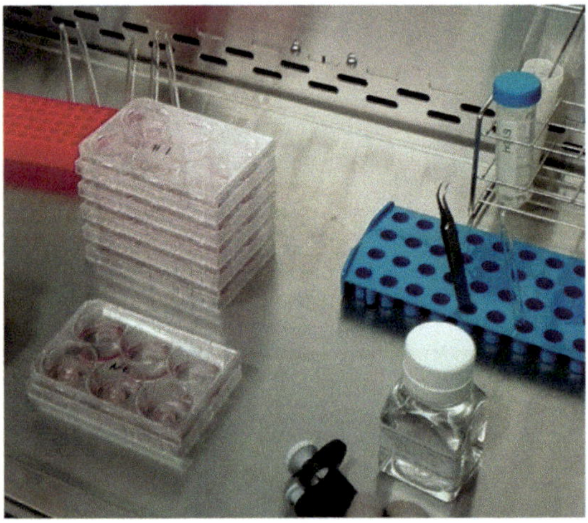

Figure 4.9 In vitro skin irritation test setup at ACEL Inc.

Table 4.4 Results of in vitro skin irritation test for ozone UFB water

Request analysis destination	ACEL Inc.	Food and Drug Safety Center
3D tissue model	EpiDerm EPI-200	LabCyte EPI-MODEL24
Manufacturer	MatTak Co. (USA)	Japan Tissue Engineering Co., Ltd.
Exposure time	60 min	15 min
Start date and time	2016/1/19 (Tue) 12:30	2016/1/20 (Wed) 14:30
Positive control (PC)	5% sodium dodecyl sulfate (SDS)	5% SDS
Negative control (NC)	D-PBS (-), Dulbecco's Phosphate Buffered Saline: Removed Mg & Ca from PBS	Phosphate Buffered Saline (PBS)
PC	Live cell rate (3.9%), irritant	Live cell rate (3.3%), irritant
NC	Live cell rate (100.0%), nonirritant	Live cell rate (100.0%), nonirritant
O_3-UFB (original)	Live cell rate (101.3%), nonirritant	Live cell rate (132.6%), nonirritant
O_3-UFB (1/4 dilution)	Live cell rate (102.2%), nonirritant	Live cell rate (129.5%), nonirritant
O_3-UFB (1/16 dilution)	Live cell rate (106.9%), nonirritant	Live cell rate (125.1%), nonirritant
0.1% NaClO	Live cell rate (104.3%), nonirritant	Live cell rate (105.2%), nonirritant

Live cell rate: ≤50% for irritant and >50% for nonirritant.

The major testing conditions and the test results are summarized in Table 4.4. The test results of both laboratories clearly indicated that the ozone bubble water was nonirritant. For verification, 0.1% sodium hypochlorite solution was used for chemical sterilization, and the results indicated that the water was nonirritant. The skin acridity of the ozone bubble water is equal to either 0.1% sodium hypochlorite or lower. Because the sterilization mechanism of ozone UFB water is an oxidation–reduction reaction, the value of the oxidation–reduction potential is regarded to be a useful index. However, the sterilization mechanism of sodium hypochlorite does not depend on the oxidation–reduction potential, and their comparison is not direct.

4.5 Conclusion

A few cases of real UFB sample measurements have been introduced in this chapter. The laser pointer method was widely employed to demonstrate the presence of UFB. The UBM apparatus, which digitizes the intensity of the scattered light, has the advantage of being portable.

Further, qNano, a sophisticated particle measurement system using the TRPS principle, was used for real UFB water generated by three different types of generators at distinct locations. The reproducibility of these measurements was not good, which indicated that the properties of UFB were influenced by the transportation and storage conditions. Furthermore, the UFB water used to irrigate lettuce at a plant factory was evaluated by qNano, in which the number concentration increased and decreased at sampling points along the irrigating system, suggesting contamination by the system itself. The human safety tests of ozone UFB water conducted at two different laboratories demonstrated that the water was nonirritant.

4.6 Appendix

4.6.1 Appendix A: UFB Obtained from Keio University

Table 4A.1 Results obtained by qNano (NP200) for the UFB water from the Keio University (2 days since bubble generation)

Run	Dilution	Particle count	Particle rate/min	Mean Dia./nm	Mode Dia./nm	Measured Conc./mL	Raw Conc./mL
1	1.0345	303	59.9	154	133	1.7E+08	1.7E+08

2		560	63.3	156	124	1.8E+08	1.8E+08
3		731	78.8	155	131	2.2E+08	2.3E+08
4		530	63.0	157	127	1.7E+08	1.8E+08
5		868	86.8	157	131	2.4E+08	2.5E+08
STD	1500	509	362.1		204	1.0E+09	1.5 E+12

*3 μL of 3.3M KCl, including 0.9% Triton-X-100, was added to the sample of 87 μL.

Table 4A.2 Results obtained by qNano (NP100) for the UFB water from the Keio University (9 days since bubble generation)

Run	Dilution	Particle count	Particle rate/ min	Mean Dia./nm	Mode Dia./nm	Measured Conc./mL	Raw Conc./mL
1		175	33.9	73	64	3.1E+08	3.2E+08
2		356	71.1	68	60	6.5E+08	6.7E+08
3	1.0345	171	34.1	89	58	3.1E+08	3.2E+08
4		83	17.0	83	60	1.5E+08	1.6E+08
5		139	27.7	72	57	2.5E+08	2.6E+08
STD	450	525	366.8		119	3.3E+09	1.5E+12

*3 μL of 3.3M KCl, including 0.9% Triton-X-100, was added to the sample of 87 μL.

Table 4A.3 Results obtained by qNano (NP200) for the UFB water from the Keio University (22 days since bubble generation)

Run	Dilution	Particle count	Particle rate/ min	Mean Dia./nm	Mode Dia./nm	Measured Conc./mL	Raw Conc./mL
1		11	0.9	217	121	1.3E+06	1.3E+06
2		15	1.5	202	117	2.1E+06	2.2E+06
3	1.0345	26	2.6	194	143	3.7E+06	3.8E+06
4		17	1.5	221	171	2.1E+06	2.2E+06
5		16	1.6	206	172	2.3E+06	2.3E+06
STD	1500	524	704.1		204	1.0E+09	1.5 E+12

*3 μL of 3.3M KCl, including 0.9% Triton-X-100, was added to the sample of 87 μL.

Table 4A.4 Results obtained by qNano (NP100) for the UFB water from the Keio University (14 days since bubble generation)

Run	Dilution	Particle count	Particle rate/ min	Mean Dia./nm	Mode Dia./nm	Measured Conc./mL	Raw Conc./mL
1	1.1111	132	13.2	60	56	6.7E+07	7.4E+07
2		122	12.2	59	58	6.2E+07	6.8E+07
3		95	14.4	64	57	7.3E+07	8.1E+07
4		108	10.8	62	59	5.4E+07	6.0E+07
5		132	11.9	60	56	6.0E+07	6.7E+07
STD	450	532	661.3		119	3.3E+09	1.5E+12

*3 μL of 3.3M KCl, including 0.9% Triton-X-100, was added to the sample of 87 μL.

4.6.2 Appendix B: UFB Obtained from Osaka University

Table 4B.1 Results obtained by qNano (NP100) for the UFB water from the Osaka University transported through plastic containers

Run	Dilution	Particle count	Particle rate/ min	Mean Dia./nm	Mode Dia./nm	Measured Conc./mL	Raw Conc./mL
1	1.1111	102	10.2	84	65	2.8E+07	3.1E+07
2		126	12.6	79	55	3.4E+07	3.8E+07
3		171	17.1	91	82	4.7E+07	5.2E+07
4		103	10.3	88	61	2.8E+07	3.1E+07
5		95	9.5	84	79	2.6E+07	2.9E+07
STD	450	553	1219.5		119	3.3E+09	1.5E+12

*10 μL of 1M KCl, including 0.3% Triton-X-100, was added to the sample of 90 μL.

Table 4B.2 Results obtained by qNano (NP100) for the UFB water from the Osaka University transported through glass bottles

Run	Dilution	Particle count	Particle rate/ min	Mean Dia./nm	Mode Dia./nm	Measured Conc./mL	Raw Conc./mL
1	1.1111	60	6.0	63	61	2.1E+07	2.3E+07
2		194	26.0	93	70	8.9E+07	9.9E+07

3		361	36.8	87	72	1.3E+08	1.4E+08
4		318	31.8	86	68	1.1E+08	1.2E+08
5		265	32.3	83	72	1.1E+08	1.2E+08
STD	450	538	973.6		119	3.3E+09	1.5E+12

*10 μL of 1M KCl, including 0.3% Triton-X-100, was added to the sample of 90 μL.

4.6.3 Appendix C: UFB Obtained from Lettuce Factory

Table 4C.1 Results obtained by qNano (NP100) for the UFB water from lettuce factory using UFB generator

Run	Dilute	Mean Dia./nm	Mode Dia./nm	Measured Conc./mL	Raw Conc./mL	Particle count	Particle rate/ min
A1		90	85	1.2E+09	1.3E+09	583	58
A2		74	67	1.2E+08	1.3E+08	57	6
A3	1.111	79	83	1.4E+08	1.6E+08	70	7
A4		80	76	2.8E+08	3.1E+08	137	14
A5		87	84	5.4E+08	6.0E+08	263	26
B1		79	72	2.3E+08	2.5E+08	111	11
B2		89	113	3.3E+09	3.7E+09	2,163	162
B3	1.111	79	84	6.8E+09	7.5E+09	3,302	330
B4		84	85	6.2E+09	6.8E+09	3,008	301
B5		79	74	3.6E+09	4.0E+09	1,765	177
C1		80	70	9.8E+08	1.1E+09	477	48
C2		82	93	2.9E+08	3.2E+08	140	14
C3	1.111	74	59	1.0E+07	1.1E+07	5	1
C4		–	–	2.0E+06	2.3E+06	1	0.1
C5		–	–	4.1E+06	4.6E+06	2	0.2
STD	450		120	3.3E+09	1.5E+12	583	163

*Lettuce Factory UFB area; A: upstream, B: middle stream, C: downstream.
All samples were treated with 0.22 μm MF.
10 μL of 1M KCl, including 0.3% Triton-X-100, was added to the sample of 90 μL.

References

1. Maas, S. L., De Vrij, J., and Broekman, M. L. (2014). Quantification and size-profiling of extracellular vesicles using tunable resistive pulse sensing. *Journal of Visualized Experiments*, **92**, e51623. https://doi.org/10.3791/51623
2. Peyman, S. A., McLaughlan, J. R., Abou-Saleh, R. H., Marston, G., Johnson, B. R. G., Freear, S., Coletta, P. L., Markham, A. F., and Evans, S. D. (2016). On-chip preparation of nanoscale contrast agents towards high-resolution ultrasound imaging. *Lab on a Chip*, **16**(4), pp. 679–687. http://dx.doi.org/10.1039/C5LC01394A.
3. Kobayashi, H., Maeda, S., Nishihara, K., Fujita, T., Maruyama, M., and Shimaoka, H. (2015). Measurement of number concentration of ultrafine bubbles using laser diffraction method and comparison with particle tracking analysis method. *Proceedings of the Multiphase Flow Symposium*, Japanese Society for Multiphase Flow, E-115 (in Japanese).
4. Sumoto, T., Maeda, H., Ogawa, T., and Akasaka, T. (2016). Concentration evaluation by quantitative laser diffraction method (qLD). *Journal of the Society of Powder Technology, Japan*, **53**(5), pp. 306–309 (in Japanese). https://doi.org/10.4164/sptj.53.306.
5. Sonoda, A. (2017). Measurement of ultra fine bubble using laser diffraction method. *Journal of the Society of Powder Technology, Japan*, **54**(9), pp. 590–595 (in Japanese). https://doi.org/10.4164/sptj.54.590.
6. Tuziuti, T., Yasui, K., and Kanematsu, W. (2017). Influence of increase in static pressure on bulk nanobubbles. *Ultrasonics Sonochemistry*, **38**, pp. 347–350. https://doi.org/10.1016/j.ultsonch.2017.03.036.
7. Tuziuti, T., Yasui, K., and Kanematsu, W. (2018). Influence of addition of degassed water on bulk nanobubbles. *Ultrasonics Sonochemistry*, **43**, pp. 272–274. https://doi.org/10.1016/j.ultsonch.2018.01.015.
8. Kinoshita, T. (2001). The method to determine the optimum refractive index parameter in the laser diffraction and scattering method. *Advanced Powder Technology*, **12**(4), pp. 589–602. https://doi.org/10.1163/15685520152756697.

Chapter 5

Theory of Ultrafine Bubbles

Kyuichi Yasui
National Institute of Advanced Industrial Science and Technology (AIST), Nagoya 463-8560, Japan
k.yasui@aist.go.jp

In this chapter, mysteries of ultrafine bubbles are discussed based on theoretical studies. One of the major mysteries is the mechanism of stability of an ultrafine bubble against dissolution. The dynamic equilibrium model is discussed in detail as the most promising hypothesis. Another mystery is the problem whether OH radicals are produced from ultrafine bubbles after stopping cavitation for the generation of fine bubbles. It is suggested that the chemical reaction of H_2O_2 and O_3 in the liquid produced from violently collapsed bubbles during cavitation results in the production of OH radicals after stopping cavitation. Finally, the effect of ultrafine bubbles on the surface tension of water is discussed.

5.1 Introduction

There are two major mysteries in ultrafine bubbles [1]. One is the mechanism for the stability of an ultrafine bubble. The other

Ultrafine Bubbles
Edited by Koichi Terasaka, Kyuichi Yasui, Wataru Kanematsu, and Nobuhiro Aya

ISBN 978-981-4877-59-6 (Hardcover), 978-1-003-14195-2 (eBook)
www.jennystanford.com

is the problem whether OH radicals are produced from ultrafine bubbles or not. In this chapter, these two problems are theoretically discussed. For the stability, the most promising hypothesis is the dynamic equilibrium model; an ultrafine bubble is stabilized against dissolution by partial coverage of its surface with a hydrophobic material [1–3]. For the OH radical production, it is suggested that the chemical reaction of H_2O_2 and O_3 in the liquid produced from violently collapsed microbubbles during hydrodynamic or acoustic cavitation used for the production of ultrafine bubbles may be the origin of the experimentally observed OH radicals after stopping cavitation [4–6]. It is numerically shown that a negligible amount of OH radicals is produced from a dissolving ultrafine bubble [7, 8].

The role of ultrafine bubbles on the surface tension of water is also discussed [9, 10]. It is suggested that ultrafine bubbles partly covered with hydrophobic materials are concentrated at the surface of water. Then, a liquid film is more easily ruptured when ultrafine bubbles burst and disappear at the liquid surface by directing the uncovered part of an ultrafine bubble into the gas phase above the liquid. It decreases the value of "surface tension" measured by the du Noüy ring method [10].

5.2 Stability

5.2.1 Reason for Skepticism

Many researchers, especially in fluid dynamics, do not believe that an ultrafine bubble is stable for a month or more [11]. Such researchers often regard the experimentally observed ultrafine bubbles as solid impurities. One of the reasons for their consideration is as follows. It is well known that the pressure inside a bubble (p_{in}) is higher than the ambient liquid pressure (p_0) by the Laplace pressure ($2\sigma/R$, where σ is surface tension and R is bubble radius) (Eq. 5.1, Fig. 5.1) [12, 13].

$$p_{in} = p_0 + \frac{2\sigma}{R} \tag{5.1}$$

The surface tension of pure water at 20°C is 7.275×10^{-2} (N/m). Thus, the Laplace pressure is 14.55 bar (= 1.455×10^6 Pa = 14.36

atm) for an ultrafine bubble of R = 100 nm (Fig. 5.2) [13]. In other words, the pressure inside a typical ultrafine bubble is more than one order of magnitude higher than the ambient liquid pressure (1 atm). As the pressure of gas dissolved in liquid is usually almost equivalent to the ambient pressure (1 atm), the gas immediately dissolves into liquid water from the interior of a bubble. The time (t_{diss}) for the complete dissolution of a bubble into gas-saturated water is estimated by the following equation derived by the Epstein–Plesset theory [14].

$$t_{\text{diss}} = \frac{1}{3}\left(1 + \frac{R_0 p_0}{2\sigma}\right)\left(\frac{R_0^2 M_{\text{gas}} p_0}{D_{\text{gas}} c_{s0} R_g T}\right) \tag{5.2}$$

where R_0 is the initial bubble radius, M_{gas} is the molar weight of the gas, D_{gas} is the diffusion coefficient of the gas in the liquid, c_{s0} is the saturated gas concentration in the liquid far from a bubble, R_g is the universal gas constant (= 8.3145 J/mol K), and T is temperature. According to Eq. 5.2, the time for the complete dissolution of a typical ultrafine bubble of R = 100 nm is about 77.8 μs (Fig. 5.3) [14]. Here the following values are used for an air bubble in water saturated with air at 20°C; p_0 = 1 atm = 1.01325 × 10^5 Pa, σ = 7.275 × 10^{-2} N/m, M_{gas} = 28.96 × 10^{-3} kg/mol, D_{gas} = 2.4 × 10^{-9} m^2/s, c_{s0} = 2.3 × 10^{-2} kg/m^3, and T = 293 K. This very short lifetime of an ultrafine bubble predicted by the Epstein–Plesset theory is the reason for the skepticism for stable ultrafine bubbles. If the liquid water is strongly oversaturated with gas (air), the lifetime of an ultrafine bubble becomes infinitely long. If the gas pressure dissolved in the liquid is as high as 15.36 atm, the lifetime of an ultrafine bubble of R = 100 nm is theoretically long. However, this equilibrium is unstable because a small increase (or decrease) in the bubble radius results in further increase (or decrease) in the bubble radius. Thus, this equilibrium state is not physically observable. This is another reason for the skepticism for stable ultrafine bubbles.

Alheshibri and Craig [15, 16] experimentally measured the mean density of an ultrafine bubble as 0.91–0.95 g/cm^3 by the resonant mass measurement varying the ratio of H_2O and D_2O to dilute a single batch of ultrafine bubbles. The density was significantly higher than that of gas. Furthermore, the application of an external pressure of 10 atm resulted in an insignificant change in the diameter of ultrafine bubbles. Thus, they concluded that "ultrafine bubbles" are not bubbles but solid or liquid particles.

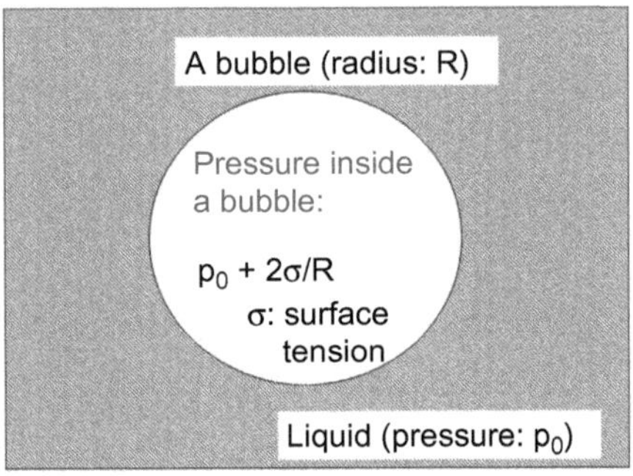

Figure 5.1 Laplace pressure. Reprinted with permission from Yasui [13]. Copyright (2016), The Japanese Society for Multiphase Flow.

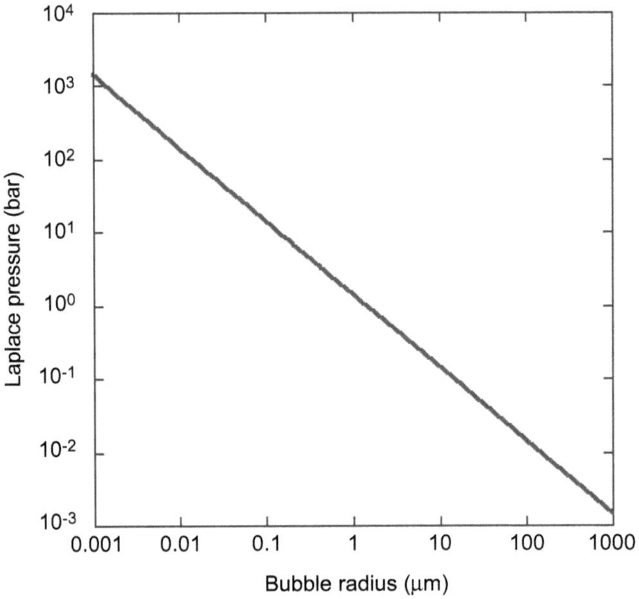

Figure 5.2 Laplace pressure as a function of bubble radius. Reprinted with permission from Yasui [13]. Copyright (2016), The Japanese Society for Multiphase Flow.

In the following subsections, several proposed models for the stability of an ultrafine bubble are discussed. In the last subsection, the dynamic equilibrium model is discussed in detail because it is the most promising one supported by transmission electron microscope (TEM) images of ultrafine bubbles without freezing [3]. The interpretation of the experimental results of Alheshibri and Craig [15, 16] is also given.

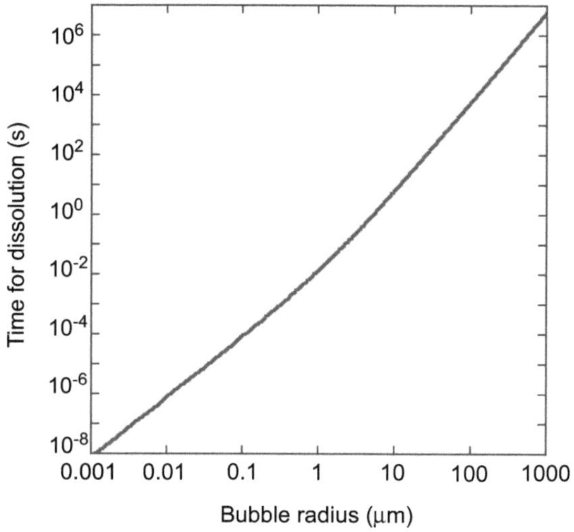

Figure 5.3 The time for complete dissolution of an air bubble in water saturated with air as a function of bubble radius calculated by the Epstein–Plesset theory. Reprinted with permission from Yasui [14]. Copyright (2015), Elsevier.

5.2.2 Electrostatic Repulsion Model

It has been experimentally reported that an ultrafine bubble is negatively charged in pure water [17–20]. The zeta potential of an ultrafine bubble as well as a microbubble is roughly about −40 mV in pure water at pH = 7 [17–20]. According to the electrostatic repulsion model, the electrostatic repulsion of the negatively charged surface of a bubble may balance with the Laplace pressure (Fig. 5.4) [12, 13]. In other words, gas diffusion from the interior of a bubble into water is stopped by the reduction in the gas pressure inside a bubble due to the electrostatic repulsive force. In this case, the pressure inside a bubble is given as follows instead of Eq. 5.1 [12, 13].

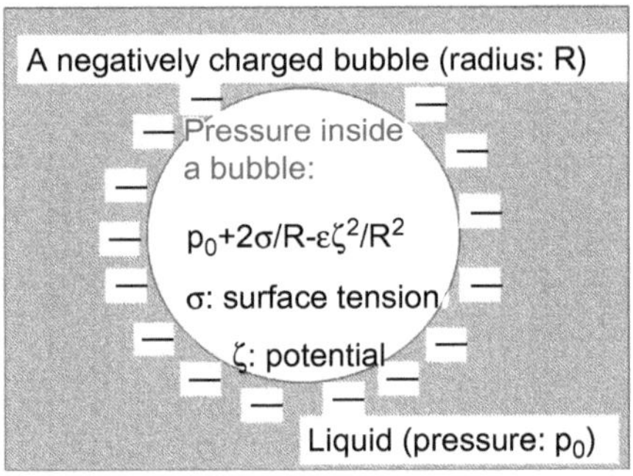

Figure 5.4 Electrostatic negative pressure. Reprinted with permission from Yasui [13]. Copyright (2016), The Japanese Society for Multiphase Flow.

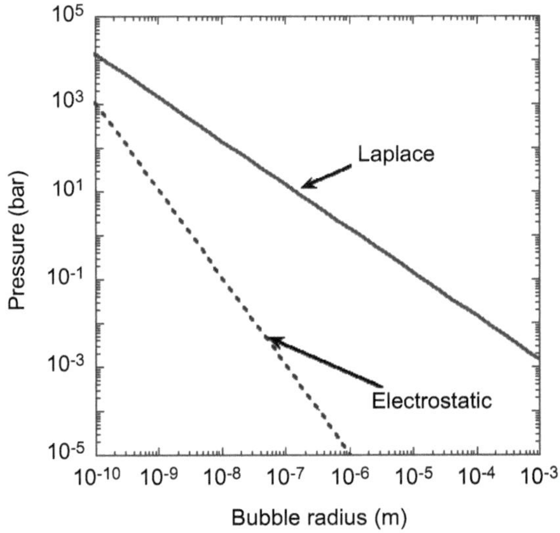

Figure 5.5 The absolute value of the electrostatic negative pressure (with $\zeta = -40$ mV) and Laplace pressure as a function of bubble radius. Reprinted with permission from Yasui [13]. Copyright (2016), The Japanese Society for Multiphase Flow.

$$p_{\text{in}} = p_0 + \frac{2\sigma}{R} - \frac{\varepsilon\zeta^2}{R^2} \tag{5.3}$$

where ε dielectric constant and ζ the zeta potential of a bubble (= –40 mV). The last term in the right side of Eq. 5.3 is the negative pressure due to the electrostatic repulsive force of the negatively charged surface of a bubble. This term is, however, more than one order of magnitude smaller than the Laplace pressure in its magnitude (Fig. 5.5) [12, 13]. It means that the electrostatic repulsive force is too weak to balance with the Laplace pressure. It seems that the model does not work for the stability of an ultrafine bubble.

5.2.3 Skin Model

When a liquid saturated with gas is irradiated with strong ultrasound, many tiny gas microbubbles are created if the pressure amplitude of ultrasound is higher than 1–2 atm [12, 14]. This is called acoustic cavitation. The threshold acoustic pressure amplitude is called cavitation threshold for acoustic cavitation. Theoretically, the cavitation threshold is estimated to be as high as 500–1000 atm by the equation of state of liquid or nucleation theory using the Boltzmann factor [21, 22]. The experimental cavitation threshold decreases from about 80 atm in highly degassed water to 1 atm in air-saturated water as the gas concentration in water increases [12, 14, 23]. The large discrepancy between the theoretical and experimental cavitation thresholds as well as decrease in the cavitation threshold by gas concentration strongly suggests that there are many cavitation nuclei in liquid water. There are two types of cavitation nuclei: solid particles and gas bubbles [24, 25]. As a candidate for the latter, the skin model has been suggested for more than 60 years [26–28]. A gas bubble is completely covered with a skin such as organic materials, surfactants, etc. (Fig. 5.6) [13]. The skin completely blocks the gas diffusion from the interior of a bubble to the surrounding liquid. Then, a gas bubble is stabilized against dissolution. When a gas bubble shrinks with (organic) materials on its surface, the fraction of surface coverage by (organic) materials increases. Finally, the bubble surface is completely (100%) covered with (organic) materials and stabilized by the skin model. This scenario is, however, actually difficult especially when (organic) materials are solid because the complete (100%) coverage of the bubble surface is practically impossible.

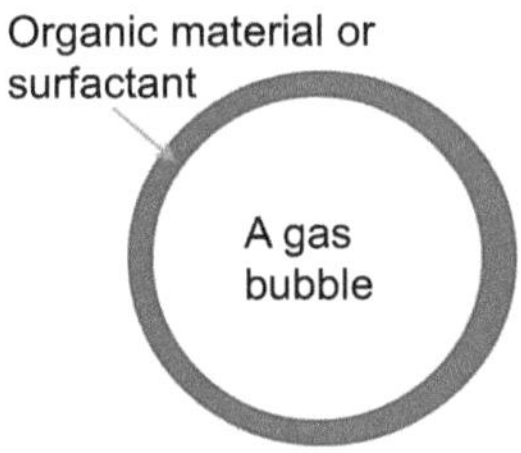

Figure 5.6 Skin model. Reprinted with permission from Yasui [13]. Copyright (2016), The Japanese Society for Multiphase Flow.

5.2.4 Particle Crevice Model

The particle crevice model has also been suggested as a model of cavitation nuclei for more than 60 years [24, 25, 29–31]. It has been experimentally reported that cavitation threshold actually decreased by the presence of solid particles [32, 33]. Gas bubbles are nucleated from a crevice on a solid particle (solid impurity) under ultrasound. A gas "bubble" is stabilized against dissolution in a crevice (Fig. 5.7) because the surface of a "bubble" is concave and surface tension *decreases* the internal pressure of the gas [1, 13]. In this mode, however, the main body is not gas but solid particle. This may contradict with the experimental results of the resonance mass measurement for ultrafine bubbles that their density is smaller than that of water.

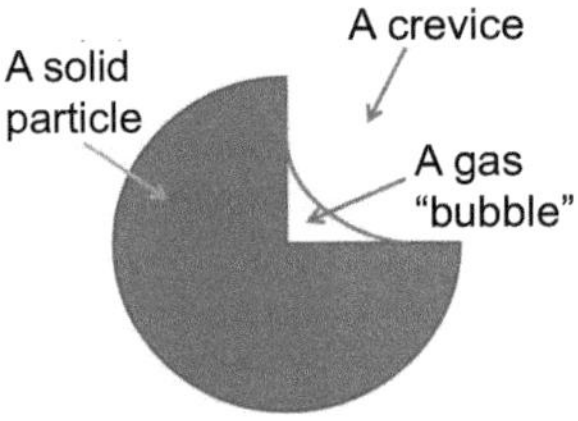

Figure 5.7 Particle crevice model. Reprinted with permission from Yasui [13]. Copyright (2016), The Japanese Society for Multiphase Flow.

5.2.5 "Armored" Bubble Model

It has been experimentally reported that a microbubble (~100 μm in diameter) is stabilized against dissolution when its surface is covered with a surfactant and spherical gold nanoparticles (~15 nm in diameter) (Fig. 5.8) [1, 34]. The lifetime of such a microbubble is more than one month. Without nanoparticles, a microbubble is stabilized only for several hours with a surfactant. The reason for the stabilization in the "armored" bubble model is the block of gas diffusion by nanoparticles and lowered Laplace pressure due to the lowered curvature of the bubble surface by the presence of nanoparticles [35]. As a model for an ultrafine bubble, however, it is not applicable in the absence of nanoparticles and surfactants.

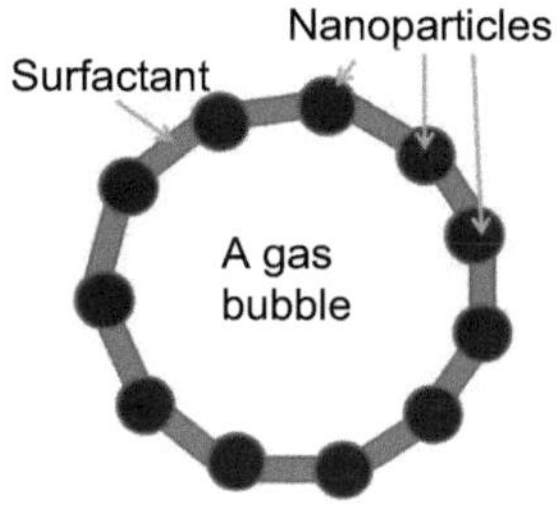

Figure 5.8 "Armored" bubble model. Reprinted with permission from Yasui et al. [1]. Copyright (2018), Elsevier.

5.2.6 Many-Body Model

As the internal pressure of an ultrafine bubble is much higher than the ambient liquid pressure, the presence of many ultrafine bubbles may keep the oversaturation of liquid with gas (Fig. 5.9) [1, 36]. If oversaturation is kept by the presence of many ultrafine bubbles, each ultrafine bubble may be stabilized against dissolution. This many-body mode is, however, not applicable to an isolated ultrafine bubble.

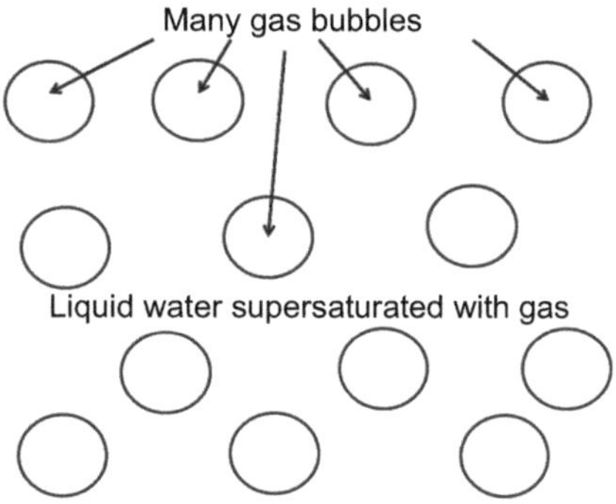

Figure 5.9 Many-body model. Reprinted with permission from Yasui et al. [1]. Copyright (2018), Elsevier.

5.2.7 Dynamic Equilibrium Model

According to the dynamic equilibrium model, an ultrafine bubble is stabilized against dissolution when more than 50% of its surface is covered with a hydrophobic material (Fig. 5.10) [2]. A hydrophobic material repels water. As a result, a depletion layer is formed on the surface of a hydrophobic material in liquid water [37, 38]. In the depletion layer, water density is decreased to 44–94% with the thickness of the layer as 0.2–5 nm, which has been experimentally measured with X-ray or neutron reflectivity measurement (Fig. 5.11) [37, 38]. In the depletion layer, gas is preferentially trapped [39–43]. As a result, the gas pressure near the surface of a hydrophobic material becomes significantly higher than the ambient gas pressure. It is crudely estimated by the following equation [2, 44, 45]:

$$p_{\text{dis}} = p_{\text{dis},\infty} e^{\frac{a}{k_B T}} \tag{5.4}$$

where p_{dis} is the pressure of gas dissolved in liquid near the surface of a hydrophobic material, $p_{\text{dis},\infty}$ is the ambient pressure of gas dissolved in liquid far from a bubble, a is the hydrophobic potential (a = 1.7 × 10^{-20} J) [44–46], k_B is the Boltzmann constant (= 1.38 × 10^{-23} J/K), and T is temperature in K. At 20°C (293 K), the pressure of gas near the surface of a hydrophobic material is as high as about 67 atm when the ambient pressure of gas dissolved in liquid is 1 atm according to Eq. 5.4.

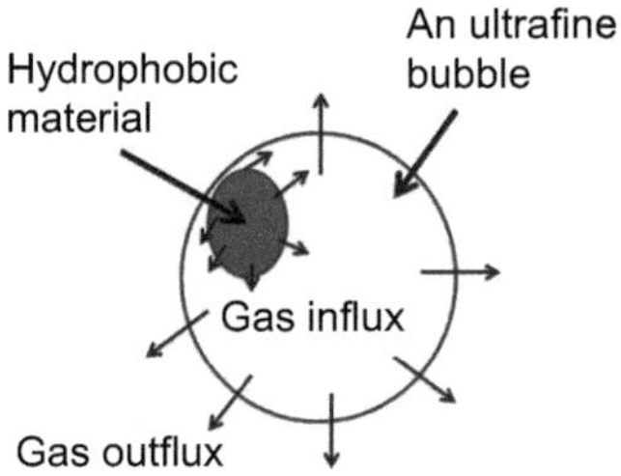

Figure 5.10 Dynamic equilibrium model. Reprinted with permission from Yasui et al. [2]. Copyright (2016), American Chemical Society.

As the pressure of gas near the surface of a hydrophobic material (67 atm) is higher than the gas pressure inside an ultrafine bubble (15 atm), gas diffuses into a bubble near the peripheral edge of the hydrophobic material on the bubble surface (Fig. 5.10) [2]. When it balances with the gas diffusion out of a bubble from the other uncovered surface of a bubble, a bubble is stabilized against dissolution. The mass balance is expressed by the following equation [2]:

$$0 = 4\pi R^2(1-r)D\frac{\rho_{H_2O}(p_{dis,\infty} - p_{gas})}{KR} + 2\pi R(\sin\theta)D\delta\frac{\rho_{H_2O}\left(p_{dis,\infty}e^{\frac{a}{k_B T}} - p_{gas}\right)}{K\delta} \tag{5.5}$$

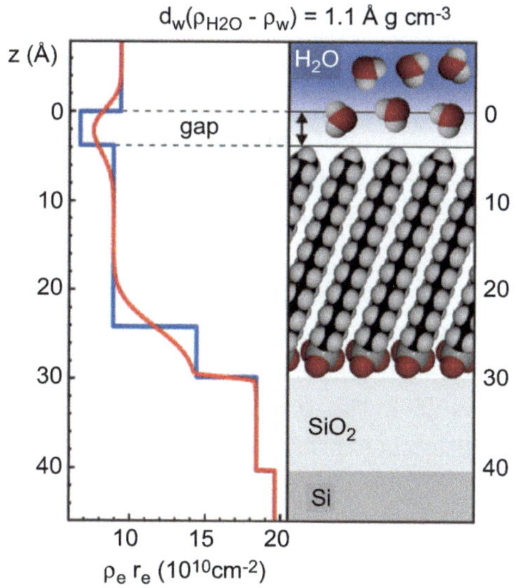

Figure 5.11 Electron density profile (hydrophobic OTS (octadecyl-trichlorosilane) immersed in water) deduced from the measured X-ray reflectivity pattern (straight blue lines: box model for the electron density, curved red lines: density profile including the roughness of the layers). A depletion layer is seen between the OTS layer and the bulk water. Reprinted with permission from Mezger et al. [38]. Copyright (2008), AIP Publishing.

where R is the bubble radius, r is the fraction of surface coverage by a hydrophobic material, $4\pi R^2$ $(1 - r)$ is the uncovered area of the bubble surface (Fig. 5.12a), D is the diffusion coefficient of gas in the liquid, ρ_{H_2O} is the molar concentration of water

(= 5.56 × 10^4 mol/m^3), K is the Henry's constant, p_{gas} is the gas pressure inside a bubble, θ is the half apex angle of the cone with its base bordered by a circumference of the covered area (Fig. 5.12a), $2\pi R$ (sin θ) is the length of the circumferential border of the hydrophobic material, and δ is the thickness of the diffusion boundary layer at the circumferential border of the hydrophobic material ($\delta << R$ is assumed) (Fig. 5.12b). The first term in the right side of Eq. 5.5 is the gas outflux. The second term is the gas influx near the peripheral edge of the hydrophobic material on the bubble surface. The gas influx is originated in the gradient in gas concentration in liquid in the boundary layer of a hydrophobic material near the bubble wall. The pressure of gas in the liquid at the bubble wall is equivalent to the gas pressure (p_{gas}) inside a bubble. In the boundary layer, the pressure of gas dissolved in liquid water is equivalent to that estimated by Eq. 5.4, apart from the bubble wall by δ. Thus, there is the gradient of gas concentration near the bubble wall in the boundary layer of a hydrophobic material (Fig. 5.12b) [2].

The gas pressure (p_{gas}) inside a bubble is calculated taking into account the effect of saturated vapor pressure (p_{vapor}) of water.

$$p_{gas} + p_{vapor} = p_0 + \frac{2\sigma}{R} \tag{5.6}$$

The mass balance is actually observable only when the equilibrium is stable. It means that a small increase (or decrease) in bubble radius results in the return of radius to the initial equilibrium value. This condition is expressed as follows [2]:

$$R_0 > \frac{\sigma f(\theta)}{\left(p_{dis,\infty} e^{\frac{a}{k_B T}} + p_{vapor} - p_0 \right)} \tag{5.7}$$

where

$$f(\theta) = 2 + \frac{(1+\cos\theta)}{\sin\theta}(1+\cos\theta+\sin\theta) \tag{5.8}$$

In the derivation of Eqs. 5.7 and 5.8, the following relationship has been used (Fig. 5.12a):

$$r = \frac{1}{2}(1-\cos\theta) \tag{5.9}$$

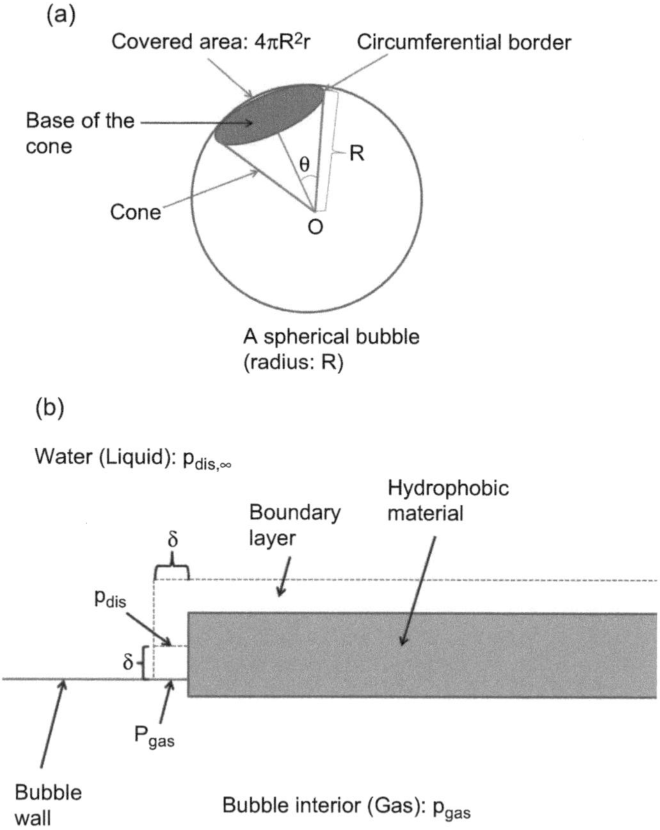

Figure 5.12 An ultrafine bubble partly covered with a piece of hydrophobic material. (a) Whole image. (b) Enlarged view at around the bubble wall partly covered with hydrophobic material. Reprinted with permission from Yasui et al. [2]. Copyright (2016), American Chemical Society.

Next, the criticism on the dynamic equilibrium model is discussed. There are mainly two reasons for the criticism. One is that the permanent gas influx and outflux assumed in the model may violate the laws of thermodynamics because it is like a perpetual motion machine. The other is that liquid flow was not experimentally detected around a surface nanobubble [47, 48]. A surface nanobubble is a stable gas object on a solid surface [49]. The dynamic equilibrium model was first proposed for the stability of a surface nanobubble, which states that the gas leaving a surface nanobubble is continuously replenished by the influx of gas near a hydrophobic solid surface [50]. The absence of the fluid flow around a surface nanobubble does not contradict with the dynamic equilibrium model because gas diffusion in quiescent liquid is assumed in the model and the liquid flow is not necessarily required [2, 45].

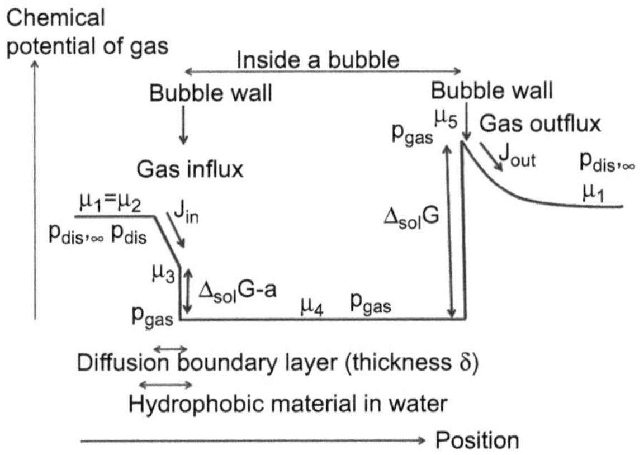

Figure 5.13 Schematic diagram of chemical potentials of gas in the dynamic equilibrium model both inside and outside a bubble partly covered with hydrophobic material. Reprinted with permission from Yasui et al. [2]. Copyright (2016), American Chemical Society.

Now the problem of a perpetual motion machine is discussed. In order to study whether the model satisfies the laws of thermodynamics, the chemical potential of gas is calculated at each step of gas diffusion (Fig. 5.13) [2]. In Fig. 5.13, $\Delta_{sol}G$ is the Gibbs energy for the dissolution of gas into liquid water. Details of the chemical potentials at each step are described in Ref. [2]. A decrease in chemical potential is observed near the bubble wall due to the gradient in gas concentration discussed previously with Fig. 5.12b. Thus, there is no problem in the diffusion of gas into a bubble. In Ref. [2], the change in energy as well as the change in entropy is calculated for each step of gas diffusion. The total change in energy and the total change in entropy are both zero. It means that energy is conserved. It satisfies the first law of thermodynamics (the energy conservation). Entropy is also conserved. The second law of thermodynamics is that the sum of the entropy changes of a system and its exterior never decreases [51]. The total entropy change is zero when the state is in equilibrium [52]. Thus, the dynamic equilibrium model satisfies the second law of thermodynamics. It is actually a kind of equilibrium state. As the dynamic equilibrium model satisfies the laws of thermodynamics, it is not a perpetual motion machine.

Numerical calculations of the mass balance condition (Eq. 5.5) and the stability condition for the mass balance (Eq. 5.7) have revealed that some ultrafine bubbles are stable when the fraction of the surface coverage by a hydrophobic material is more than 0.5 (50%) [2]. This is different from the skin model discussed in Section 5.2.3 for which a fraction of surface coverage is always 1 (100%). The stable bubble size is determined by the area of a hydrophobic material covering the surface of an ultrafine bubble (Fig. 5.14) [2]. Only the restricted range of the area of a hydrophobic material results in a stable ultrafine bubble. Correspondingly, the range of stable size for an ultrafine bubble is also restricted under a given condition such as the degree of gas saturation in the liquid water, liquid temperature, etc. For a slightly degassed water (80% of the gas saturation in Fig. 5.14), some microbubbles are also stabilized by partially covered with a hydrophobic material [2].

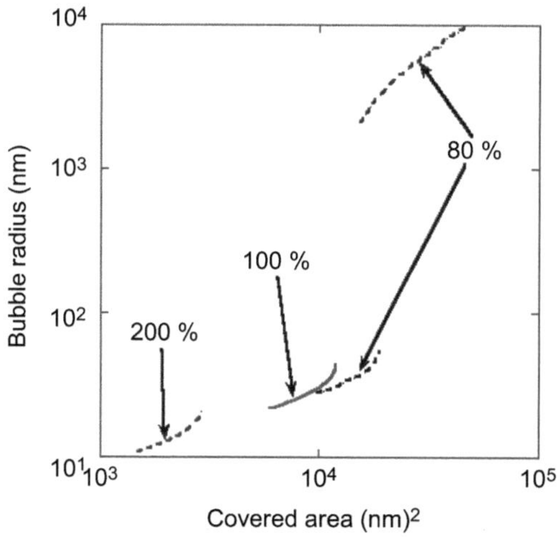

Figure 5.14 Stable bubble radius as a function of the area covered with a piece of hydrophobic material for various degrees of gas saturation of liquid water (80, 100, and 200%). The ambient liquid pressure and temperature are 1 atm and 20°C, respectively. Reprinted with permission from Yasui et al. [2]. Copyright (2016), American Chemical Society.

As the experimental evidence for the dynamic equilibrium model, Sugano, Miyoshi, and Inazato [3] reported the TEM images of stable ultrafine bubbles partly covered with hydrophobic materials in aqueous solutions without freezing. Details are described in Chapter 6 by Sugano et al.

Next, the experimental results by Alheshibri and Craig [15, 16] are discussed based on the dynamic equilibrium model. The experimentally measured density of an ultrafine bubble as 0.91–0.95 g/cm^3 could be due to a hydrophobic material covering the bubble surface. The mass of an ultrafine bubble is almost equivalent to the mass of a hydrophobic material covering the bubble surface. The insignificant change in the diameter of an ultrafine bubble by the application of external pressure of 10 atm observed experimentally could be due to the supply of the gas into a bubble by the hydrophobic material covering the bubble surface. Further studies are required on these topics.

5.3 Radical Formation

5.3.1 Experimental Results

In this section, the problem whether OH radicals are produced from ultrafine bubbles is discussed. First, the experimental reports are reviewed. Takahashi et al. [5] reported the electron spin resonance (ESR) signal of OH radicals after stopping cavitation from the liquid water in which microbubbles were generated by using hydrodynamic cavitation. Here, hydrodynamic cavitation occurred at a nozzle from which high-pressure liquid water containing high-pressure air was injected. Hydrodynamic cavitation is the formation and subsequent collapse of microbubbles [25]. The liquid sample was taken from the water after it returned to be transparent by stopping the hydrodynamic cavitation during which the water had a milky appearance [5]. In the measurements [5], various acids such as sulfuric acid (H_2SO_4), hydrochloric acid (HCl), and nitric acid (HNO_3) were added to the liquid as well as a spin-trap agent (DMPO: 5,5-dimethyl-1-pyroroline-N-oxide). From the acid solutions, the ESR spectra exhibited the four characteristic lines from DMPO-OH. It suggests that OH radicals were produced after stopping cavitation in the acid solutions. When NaCl was added instead of the acids, much weaker lines of DMPO-OH were observed.

Liu et al. [6] reported the experimental results of fluorescence spectrophotometry from water containing oxygen ultrafine bubbles. They concluded that most of the fluorescence signal observed was due to OH radicals produced from ultrafine bubbles after stopping cavitation. In the experiment, pure oxygen was introduced into a 2 L phosphate buffer (0.1 M, pH 7.4) and circulated through a generator of ultrafine bubbles using hydrodynamic cavitation for 1 h at a constant temperature of 20°C to obtain oxygen ultrafine bubble (UFB) water. The DO concentration of oxygen UFB water was above 40 mg/L. A sensitive fluorescent probe (reagent), 3′-*p*-(aminophenyl) fluorescein (APF) (5 mM) of 2 μL was added to 10 mL of water containing UFB and the fluorescence intensities were measured. This was repeated more than five times. The final concentration of APF was 1 μM. The results of the fluorescence spectrophotometry are shown in Fig. 5.15 [6]. Before ferrous ions were added, the fluorescence intensities of oxygen UFB (NB: nanobubble) water were statistically higher than those of control water (baseline fluorescence by probe). The increase in fluorescence of the APF solution caused by oxygen UFB water was approximately 40 (Fig. 5.15) [6]. This oxidative capacity of oxygen UFB water was equivalent to 0.5 mM H_2O_2 or 0.25 μM OH. According to the Fenton reaction (Eq. 5.10), ferrous iron (II) is oxidized by H_2O_2 into ferric iron (III), a hydroxyl anion, and OH radical.

$$Fe^{2+} + H_2O_2 \rightarrow Fe^{3+} + OH^- + OH \qquad (5.10)$$

In the presence of excess amounts of H_2O_2, the concentration of added ferrous ions is equal to that of OH. If the fluorescence intensity from oxygen UFB water was due to 0.5 mM H_2O_2, the addition of 20 μM ferrous ions should result in 20 μM OH radicals. Liu et al. [6] suggested that in that case the fluorescence intensity should increase by more than 2000 by the addition of 20 μM ferrous ions to oxygen UFB water. However, the actual increase was about 70. From this discussion, Liu et al. [6] concluded that oxygen UFB water was not 0.5 mM H_2O_2 but 0.25 μM OH.

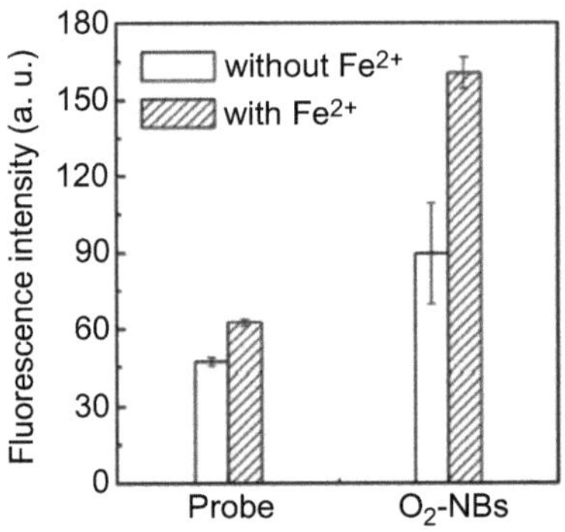

Figure 5.15 Effect of ferrous ion (20 μM) on the fluorescence intensities of the buffer solutions with and without ultrafine bubbles (NBs: nanobubbles). Reprinted with permission from Liu et al. [6]. Copyright (2016), American Chemical Society.

There is also a negative experimental report on the production of OH radicals after stopping cavitation. Tada et al. [53] reported that there was weak ESR signal of DMPO-OH from air UFB water after stopping cavitation. However, there was also weak ESR signal of DMPO-OH from pure water. The subtraction of the ESR signal of pure water from that of air UFB water results in no signal of DMPO-OH. Thus, they concluded that there was no OH radical production from air UFB water after stopping cavitation. In their experiment [53], no acid was added in contrast to the experiment of Takahashi et al. [5].

5.3.2 Radical Formation during Cavitation

It is widely known that OH radicals are produced during acoustic or hydrodynamic cavitation [12, 54]. Acoustic cavitation is the formation and subsequent collapse of micrometer-sized bubbles in liquid irradiated with strong ultrasound [12]. Ultrasound is inaudible sound with frequencies higher than 20 kHz (or sometimes 10 kHz for convenience). Sound is a propagation of pressure oscillation with a speed of sound. During the rarefaction phase of strong ultrasound, the local instantaneous pressure substantially decreases. As a result, gases dissolved in the liquid appear as gas microbubbles. The bubbles expand during the rarefaction phase of ultrasound to about 100 μm in diameter. At the compression phase of ultrasound, such bubbles violently collapse. Then, the temperature and pressure inside a bubble significantly increase at the final moment of the bubble collapse, which is called Rayleigh collapse [12, 55]. Due to the high temperature and pressure, water vapor and oxygen, if present, are dissociated inside a bubble. Consequently, oxidants such as OH radicals, H_2O_2, O_3, and O radicals are produced inside a bubble. They diffuse out of a bubble into the surrounding liquid. If there are some solutes in the liquid, the oxidants react with solutes. Such chemical reactions are called sonochemical reactions [56].

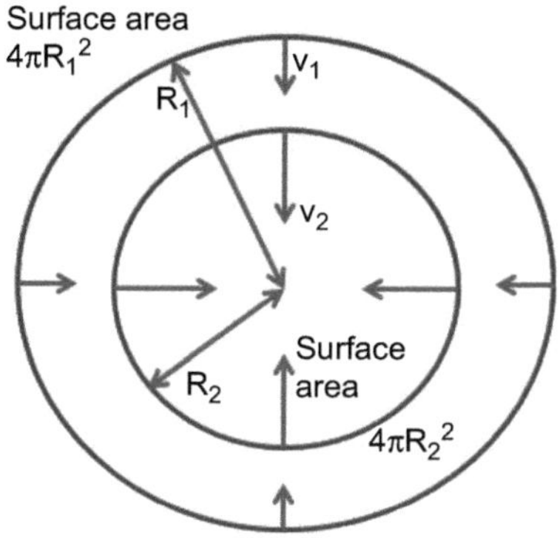

Figure 5.16 Spherically inward flow as the mechanism for the violent collapse of a bubble in acoustic or hydrodynamic cavitation. Reprinted with permission from Yasui [14]. Copyright (2015), Elsevier.

Hydrodynamic cavitation occurs in liquid associated with high-speed liquid flow such as that in a Venturi tube, swirling flow around a screw, etc. [25]. Hydrodynamic cavitation around a ship propeller is a widely known phenomenon. Erosion of a ship propeller is mostly caused by cavitation, which is called cavitation damage (or erosion). In such high-speed liquid flow, local instantaneous pressure is substantially reduced, and gas (and vapor) microbubbles appear. The bubbles expand due to the reduction in local pressure and subsequently collapse violently when the local instantaneous pressure recovers. The violent collapse of a bubble in hydrodynamic cavitation is similar to that in acoustic cavitation because the physics of the violent collapse is the same (Fig. 5.16) [12, 14, 55, 57, 58]. The reasons for the violent collapse for both acoustic and hydrodynamic cavitation [12, 14] are as follows. One is the inertia of the surrounding liquid flowing toward a bubble. The other is the spherical geometry of a collapsing bubble as a result of which the speed of the bubble collapse increases as the surface area of a shrinking bubble decreases, resulting from the conservation of mass (liquid) (Fig. 5.16) [14]. Consequently, the oxidants such as OH radicals, H_2O_2, O_3, and O radicals are produced inside a bubble also in hydrodynamic cavitation [57–59]. The cavitation damage is mostly caused by liquid jets impacting on a solid surface from collapsing cavitation bubbles near a solid wall [25].

The typical average concentration of the oxidants produced in acoustic cavitation is about 10 μM (= 10^{-5} mol/L) [12, 60]. In actual acoustic cavitation, it is difficult to compare the results of experiments and those of numerical simulations of radical formation from cavitation bubbles because there are some complexity in the phenomena such as bubble–bubble interaction, attenuation, and reflection of acoustic wave by bubbles, nonlinear propagation of acoustic wave, variation of liquid temperature, etc. [61]. For a single-bubble system shown in Fig. 5.17 [62], it is possible to directly compare the results of experiment and those of numerical simulation because a single bubble is under controlled acoustic pressure and liquid temperature. In the single-bubble system, a single stable bubble is trapped near the pressure antinode of a standing-wave field of ultrasound in the liquid partially degassed (without pre-degassing of the liquid, many cavitation bubbles appear by the irradiation of ultrasound) [55]. Under some conditions, a single stable bubble repeatedly emits faint light at each violent collapse in a clocklike manner, which is called single-bubble sonoluminescence (SBSL) [55, 62–64]. SBSL has been intensively studied since its rediscovery by Gaitan and Crum [65] in 1989. (The first experimental report on SBSL was published in 1962 [66].) Details of the researches on SBSL are reviewed in Refs. [55, 62–64]. In 2002, Didenko and Suslick [67] first reported the quantitative experimental results of single-bubble sonochemistry. The production rate of OH radicals was reported as 8.2×10^5 molecules per acoustic cycle using terephthalate dosimetry from an SBSL bubble in water at 3°C irradiated by ultrasound of 52 kHz and 1.5 atm (= 1.52 bar) in frequency and pressure amplitude, respectively. The maximum bubble radius during bubble pulsation was measured as 30.5 μm. The rate of NO_2^- ion production was measured as 9.9×10^6 ions per acoustic cycle, and the number of photons emitted in sonoluminescence was measured as 7.5×10^4 [67].

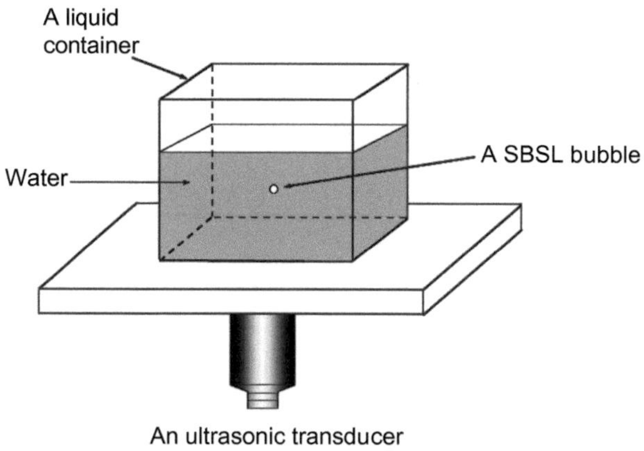

Figure 5.17 An apparatus for a single-bubble sonoluminescence (SBSL) experiment. Reprinted with permission from Yasui et al. [62]. Copyright (2004), Taylor & Francis.

Next, the results of numerical simulations are reviewed under the experimental condition of single-bubble sonochemistry. In the numerical simulations [68], the following effects have been taken into account: non-equilibrium evaporation and condensation of water vapor at the bubble wall, chemical reactions of gases and vapor inside a bubble, thermal conduction both inside and outside a bubble, variation of liquid temperature at the bubble wall, and diffusion of gases and chemical products into liquid water from the interior of a bubble. Details of the theoretical model are described in Refs. [12, 68–71]. To reproduce the experimentally observed maximum radius of a bubble (30.5 μm), the ambient bubble radius was determined as 3.6 μm. During the rarefaction phase of ultrasound, a bubble expands to 30.5 μm in radius (Fig. 5.18a) [68]. At the compression phase of ultrasound, a bubble violently collapses (Rayleigh collapse explained in Fig. 5.16). The dissolution of OH radicals into the surrounding liquid water was also numerically simulated. In Fig. 5.18b, OH flux (solid line) and its time integral (dotted line) are shown as a function of time [68]. OH flux is shown with logarithmic axis. The amount of OH radicals dissolved into the surrounding liquid water is calculated as 6.6×10^5 in the number of molecules per acoustic cycle [68]. It roughly agrees with the experimental data of 8.2×10^5 [67]. Thus, the theoretical model is almost validated.

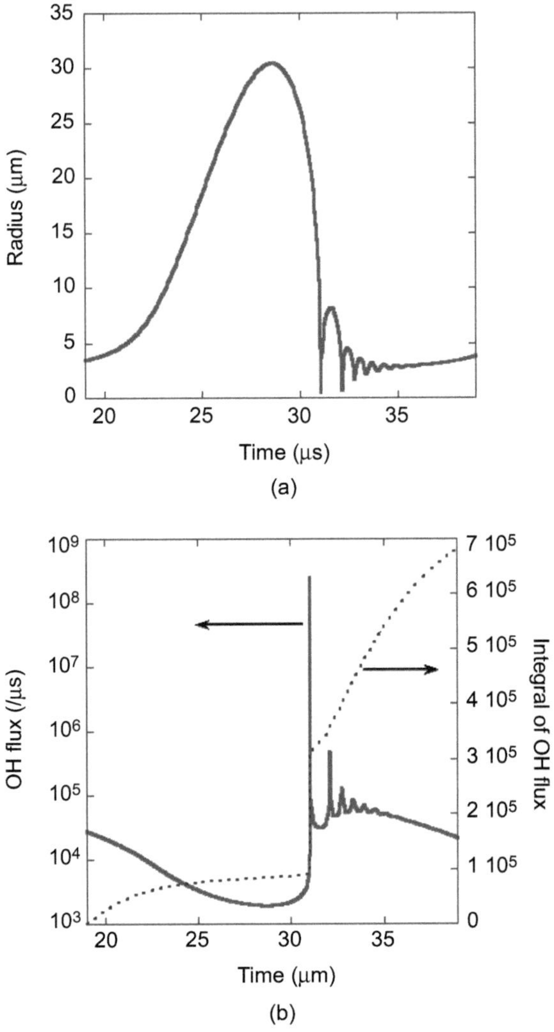

Figure 5.18 The results of numerical simulation for one acoustic cycle when an SBSL bubble in a steady state in water at 3°C is irradiated by an ultrasonic wave of 52 kHz and 1.52 bar in frequency and pressure amplitude, respectively. The ambient bubble radius is 3.6 μm. (a) The bubble radius. (b) The dissolution rate of OH radicals into the liquid from the interior of the bubble (solid line) and its time integral (dotted line). Reprinted with permission from Yasui et al. [68]. Copyright (2005), AIP Publishing.

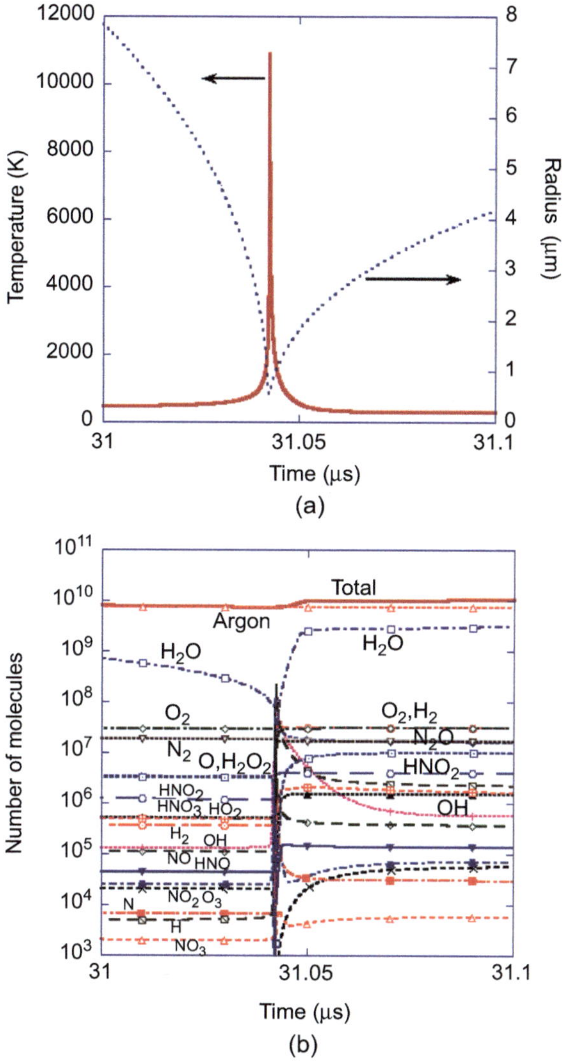

Figure 5.19 The results of numerical simulation for an SBSL bubble in a steady state at around the end of the bubble collapse only for 0.1 μs. (a) The bubble radius and the temperature inside a bubble. (b) The number of molecules inside a bubble. Reprinted with permission from Yasui et al. [68]. Copyright (2005), AIP Publishing.

The temperature inside a bubble increases up to 10,900 K at the end of the bubble collapse according to the numerical simulation

(Fig. 5.19a) [68]. As a result, many chemical species are dissociated inside a bubble (Fig. 5.19b) [68]. Most of the chemical species are created at the previous violent collapse of the stably pulsating bubble. The major non-condensable gas inside an SBSL bubble is argon according to the argon rectification hypothesis [55, 72]. The hypothesis is that nitrogen is burned inside a heated bubble and that they are changed to NO_x and HNO_x. As they are soluble species, they gradually dissolve into liquid water. Finally, only argon, which is a non-reactive species in air, remains inside an SBSL bubble (air consists of 78% nitrogen, 21% oxygen, and 1% argon in molar fraction). The argon rectification hypothesis has been validated through experimental and theoretical studies of SBSL [55]. In Fig. 5.19b, many chemical species are formed again inside a bubble after the end of the bubble collapse (after the minimum bubble radius) such as H_2, O, H_2O_2, HNO_2, and OH [68].

The amount of chemical products that dissolve into the liquid water from the interior of an SBSL bubble per acoustic cycle is listed in Table 5.1 according to the numerical simulation [68]. The dominant products are H_2, O, H_2O_2, H, HNO_2, HO_2, HNO_3, and OH. The dominant oxidants are O, H_2O_2, and OH. A small amount of O_3 is also created. The amount of HNO_2 of 2.3×10^6 is considerably lower than the experimental value for NO_2^- ions (9.9×10^6) [67, 68]. Further studies are required on this topic.

Table 5.1 Amount of chemical products that dissolve into liquid water from the interior of an SBSL bubble in a steady state per acoustic cycle according to the numerical simulation

Chemical species	Number of molecules per acoustic cycle
H_2	3.1×10^7
O	1.3×10^7
H_2O_2	6.3×10^6
H	4.1×10^6
HNO_2	2.3×10^6
HO_2	1.1×10^6
HNO_3	8.4×10^5
OH	6.6×10^5

Chemical species	Number of molecules per acoustic cycle
NO	2.5×10^5
HNO	9.5×10^4
NO_2	4.4×10^4
O_3	3.4×10^4
N	2.9×10^4
NO_3	3.1×10^3
N_2O	3.1×10^2

Source: Reprinted with permission from Yasui et al. [68]. Copyright (2005), AIP Publishing.

Table 5.2 Amount of chemical products that dissolve into liquid water from the interior of an initial air bubble in one acoustic cycle according to the numerical simulation

Chemical species	Number of molecules per acoustic cycle
HNO_2	4.0×10^7
HNO_3	3.7×10^7
O	1.6×10^7
H_2O_2	5.1×10^6
O_3	2.7×10^6
HO_2	2.3×10^6
NO_3	1.1×10^6
H_2	1.0×10^6
OH	9.9×10^5
NO_2	3.9×10^5
N_2O	3.0×10^5
NO	1.3×10^5
H	1.1×10^5
HNO	2.8×10^4
N	2.7×10^3
N_2O_5	6.8×10^2

Source: Reprinted with permission from Yasui et al. [68]. Copyright (2005), AIP Publishing.

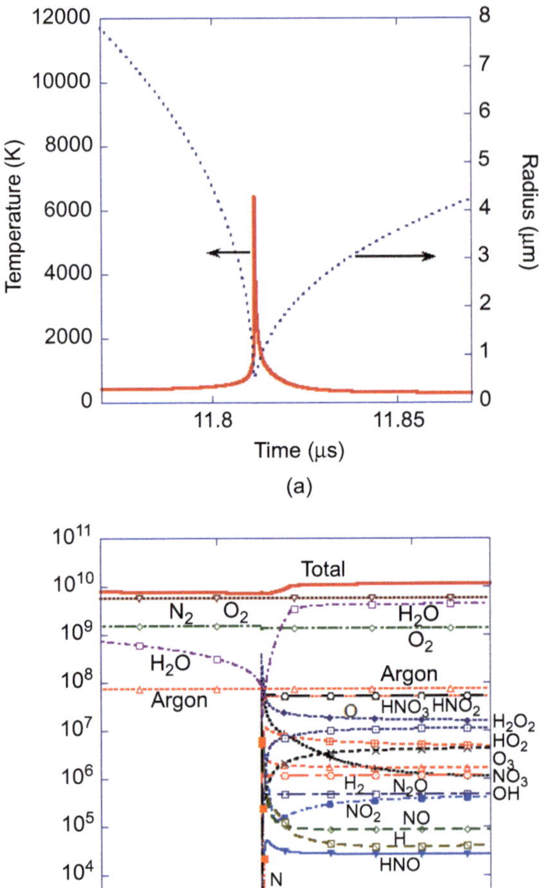

Figure 5.20 The results of numerical simulation for an initial air bubble at around the end of the bubble collapse only for 0.1 μs. (a) The bubble radius and temperature inside a bubble. (b) The number of molecules inside a bubble. Reprinted with permission from Yasui et al. [68]. Copyright (2005), AIP Publishing.

Initially, a bubble consists of air and water vapor before the establishment of the argon rectification. It needs about 100 acoustic cycles before the main bubble content completely changes from air to argon [73]. The results of the numerical simulation for the initial air bubble are shown in Fig. 5.20 [68]. The maximum bubble temperature in this case is 6500 K, which is considerably lower than that inside an argon bubble (10,900 K) (Fig. 5.20a) [68]. This is because the molar heat of air is much larger than that of argon. The main chemical products in an air bubble are HNO_2, HNO_3, O, H_2O_2, O_3, HO_2, NO_3, H_2, and OH (Fig. 5.20b and Table 5.2) [68]. The main oxidants in this case are O, H_2O_2, O_3, and OH.

5.3.3 Radical Formation in a Dissolving Bubble

After stopping cavitation, most of microbubbles move upward by buoyancy and disappear at the liquid surface as bubbles burst there [5]. Some of smaller bubbles completely dissolve into the liquid before reaching the liquid surface. Some other small bubbles are stabilized against dissolution by being partly covered with hydrophobic materials (impurities) such as oils and carbon particles, as discussed in the previous section, which are stable ultrafine bubbles [2]. In the present subsection, results of numerical simulations for a dissolving bubble are reviewed [7, 8]. Moreover, the possibility of radical formation inside a dissolving bubble is discussed based on the results of the numerical simulations.

First, the theoretical model for the simulation is briefly explained. The model is essentially the same as that used for SBSL in the previous subsection. The main difference of the present model from that for SBSL is the absence of ultrasound as well as the presence of the gas diffusion from the interior of a bubble into the surrounding liquid by Eq. 5.11 [7, 8].

$$\frac{dn_i}{dt} = -4\pi R^2 D_i \frac{(c_{s,i} - c_{\infty,i})}{R} \qquad (5.11)$$

where n_i is the number of molecules inside a bubble for the gas species i (i = N_2, O_2, or Ar), t is time, R is the instantaneous bubble radius, D_i is the diffusion coefficient of the gas species i in the liquid, $c_{s,i}$ is the saturated concentration of the species i in the liquid at the bubble wall, and $c_{\infty,i}$ is the concentration of the gas species i in the liquid far from a bubble. Here, the bubble surface is assumed as clean without any attachment of hydrophobic or hydrophilic materials. Other details of the model are described in Refs. [7, 8].

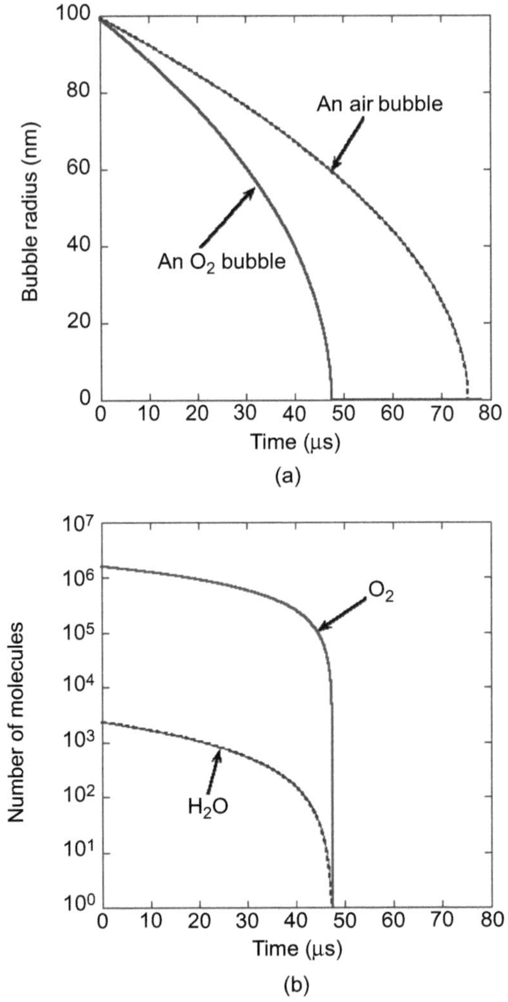

Figure 5.21 The results of numerical simulation for the dissolution of an oxygen (O_2) ultrafine bubble into water saturated with oxygen (O_2) as a function of time. The initial bubble radius is 100 nm. (a) The bubble radius. The result for an air ultrafine bubble is also shown with dotted line. (b) The number of molecules inside a bubble with logarithmic vertical axis. Reprinted with permission from Yasui et al. [8]. Copyright (2019), Elsevier.

Next, the results of the numerical simulations are described. In the present numerical simulations, the initial bubble radius is assumed as 100 nm (= 0.1 μm) because a bubble of any radius larger than 100 nm inevitably passes through the radius of 100 nm during its dissolution into liquid. Due to the larger solubility of oxygen gas in liquid water, an oxygen bubble dissolves faster than an air bubble (Fig. 5.21a) [8]. As the internal gas pressure is one order of magnitude higher than the ambient liquid pressure due to surface tension for a bubble of 100 nm in radius, the molar fraction of water vapor inside a bubble is negligibly small as the saturated vapor pressure at 20°C is only 2.3 kPa (0.0227 atm) (Fig. 5.21b) [8].

Surprisingly, the temperature inside an oxygen bubble significantly increases to 2800 K at the final moment of the bubble dissolution (Fig. 5.22a) [8]. It is slightly lower than that (3000 K) inside an air bubble due to the larger thermal conductivity of oxygen than that of air [7, 8]. The reason for the temperature increase is the same as that in the violent collapse of a bubble under ultrasound. The pV work done on a bubble by the surrounding liquid overwhelms the energy loss due to thermal conduction from the heated interior of a bubble into the surrounding liquid (Fig. 5.22b) [8].

The pressure inside a bubble increases to about 4.5 GPa at the final moment of the bubble dissolution (Fig. 5.22c) [8]. It is slightly lower than that (5 GPa) inside an air bubble due to the lower temperature of an oxygen bubble. The liquid pressure at the wall of an oxygen bubble increases to 4 GPa. The liquid temperature at the bubble wall increases to 94°C (Fig. 5.22d) [8]. However, the liquid temperature is too low for the dissociation of water molecules. The only possibility for the radical production is inside a bubble.

According to the numerical calculations of the rate of O_2 dissociation ($O_2 + O_2 \rightarrow O + O + O_2$), the total number of O atoms produced inside a dissolving oxygen bubble is on the order of 10^{-7} (Fig. 5.22e) [8]. In other words, a few O atoms may be created per 10^7 dissolving oxygen bubbles. As the typical number concentration of ultrafine bubbles is of the order of 10^8 per mL, the production of O atoms as well as OH radicals by the reaction $O + H_2O \rightarrow 2OH$ from dissolving oxygen bubbles is negligible.

It should be noted, however, that the results of the present numerical simulations are not accurate as the continuum model assumed in the present study is no longer valid at the final moment of the bubble dissolution due to the considerable increase in the Knudsen number defined by λ/R, where λ is the instantaneous mean free path of a gas molecule inside a bubble, and R is the instantaneous bubble radius (Fig. 5.22f) [8]. In the future, molecular dynamics simulations are required on the dissolution of ultrafine bubbles.

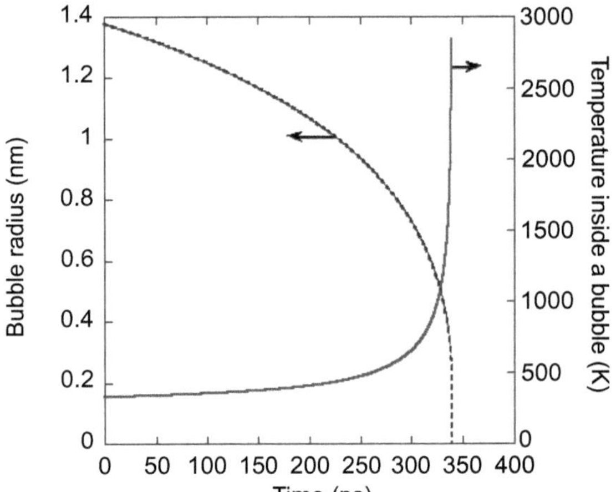

(a)

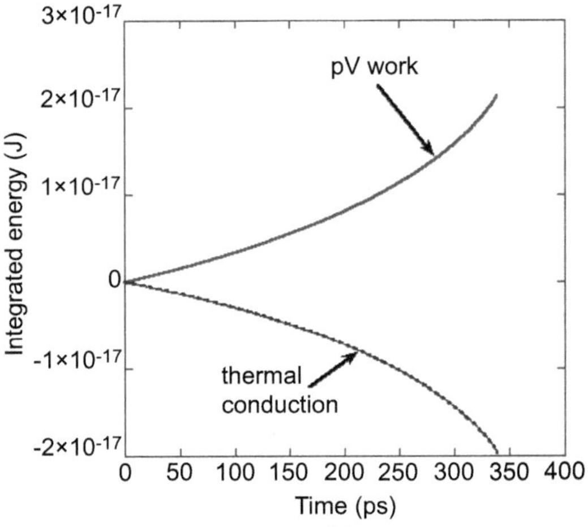

(b)

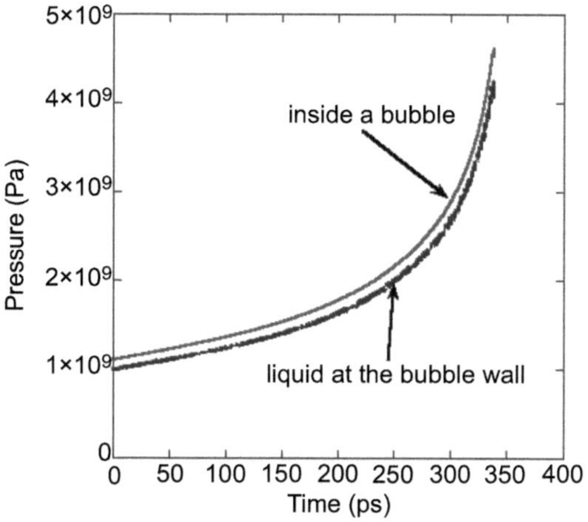

5×10⁹
4×10⁹
3×10⁹
2×10⁹
1×10⁹
0
Pressure (Pa)
inside a bubble
liquid at the bubble wall
0 50 100 150 200 250 300 350 400
Time (ps)

(c)

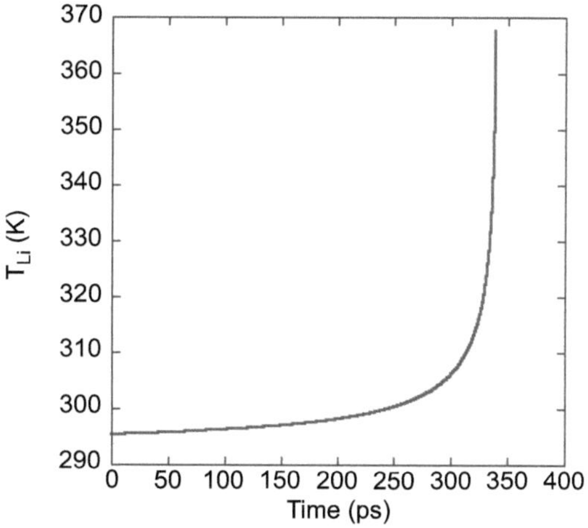

370
360
350
340
330
320
310
300
290
T_Li (K)
0 50 100 150 200 250 300 350 400
Time (ps)

(d)

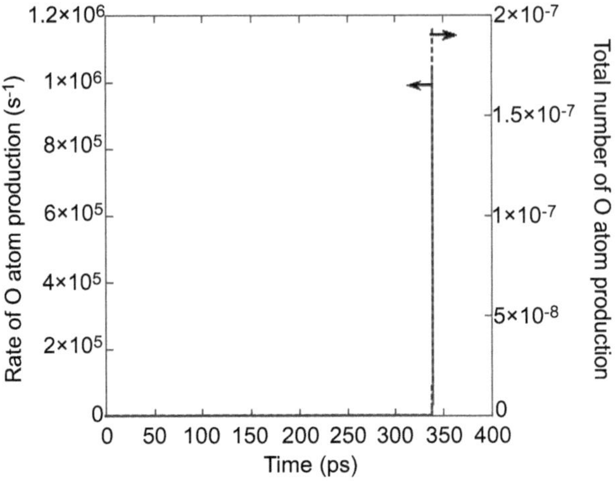

(e)

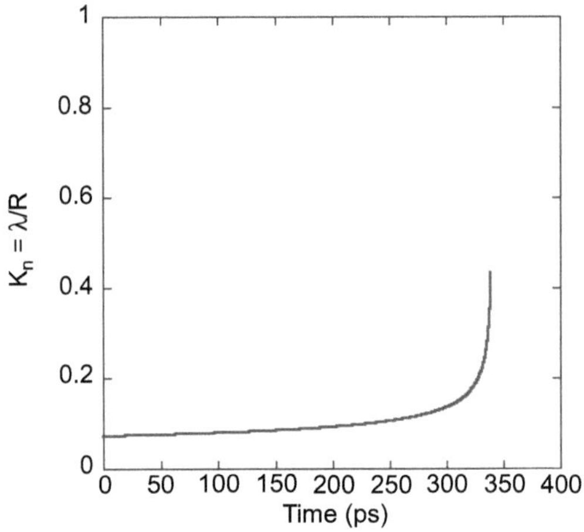

(f)

Figure 5.22 The result of numerical simulation as a function of time for the last 340 ps of the complete dissolution of an oxygen (O_2) ultrafine bubble into water. The condition is the same as that in Fig. 5.21 (t = 0 in this figure corresponds to t = 47.579 μs in Fig. 5.21). (a) The bubble radius (dotted line) and the temperature inside a bubble (solid line). (b) Time-integrated pV work done on a bubble by the surrounding liquid, and energy loss due to thermal conduction. (c) The pressure inside a bubble (solid line) and the liquid pressure at the bubble wall (dotted line). (d) The liquid temperature at the bubble wall. (e) The rate of O atom production, and the total number of O atoms produced. (f) The Knudsen number (K_n). Reprinted with permission from Yasui et al. [8]. Copyright (2019), Elsevier.

5.3.4 Radical Formation after Stopping Cavitation

In the previous subsection, it is shown that the production of OH radicals from dissolving bubbles without any dynamic stimuli is negligible [7, 8]. Furthermore, OH radicals produced during cavitation could not considerably contribute to the experimentally observed signals of OH radicals after stopping cavitation because the typical lifetime of OH radicals is only 20 ns [12]. Then, what is the mechanism of OH production after stopping cavitation experimentally reported? According to Table 5.2 [68], an appreciable amount of H_2O_2 and O_3 is produced in liquid water from an air bubble. It is known that OH radicals are produced by the reaction of H_2O_2 with O_3 in liquid water [74, 75]. When the pH of liquid water is in the range of $5 < pH < 8$, the following reactions occur [74]:

$$H_2O_2 \rightarrow HO_2^- + H^+ \qquad (5.12)$$

$$O_3 + HO_2^- \rightarrow OH + O_2^- + O_2 \qquad (5.13)$$

In acidic aqueous solutions ($pH < 5$), OH radical production is through the following slow reaction [74, 75]:

$$H_2O_2 + O_3 \rightarrow OH + HO_2 + O_2 \qquad (5.14)$$

Takahashi et al. [5] experimentally reported that the strong signal of OH radicals was detected in the ESR spectra in acidic solutions after stopping cavitation. It may be due to the slow reaction of Eq. 5.14 because OH production considerably continues even after stopping cavitation. Further studies are required on this topic.

5.4 Surface Tension

5.4.1 Experimental Results

The surface tension of liquid water containing ultrafine bubbles was measured by the du Noüy ring method by Ushida et al. [10] as about 63 mN/m, which is about 14% lower than that of pure water (about 73 mN/m) (Fig. 5.23) [9]. In an anionic surfactant (laurylbenzene sulfonic acid sodium salt: LAS) solution, there was about 20% decrease in surface tension by the presence of ultrafine bubbles for the concentration range of the surfactant up to about 500 ppm [10]. On the other hand, there is no considerable change in surface tension by the presence of ultrafine bubbles in a cationic surfactant (benzalkonium chloride: BC) solution and in a nonionic surfactant (polyoxyethylene (23) lauryl ether: AE) solution. They also experimentally studied the washing rate of an artificially soiled cloth in the presence and absence of ultrafine bubbles [10]. The washing rate was measured by the optical reflectivity of the washed cloth by using a handy colorimeter. The washing rate was determined by the optical reflectivity before and after washing as well as that for the white cloth before soiling. The definition of the washing rate is given in Ref. [10]. They reported that the washing rate considerably increased by the presence of ultrafine bubbles in the absence of surfactant. In the anionic surfactant solution, the washing rate considerably increased by the presence of ultrafine bubbles in the concentration range of surfactant up to about 700 ppm. On the other hand, there was no considerable change in the washing rate by the presence of ultrafine bubbles in the cationic surfactant solution and in the nonionic surfactant solution [10].

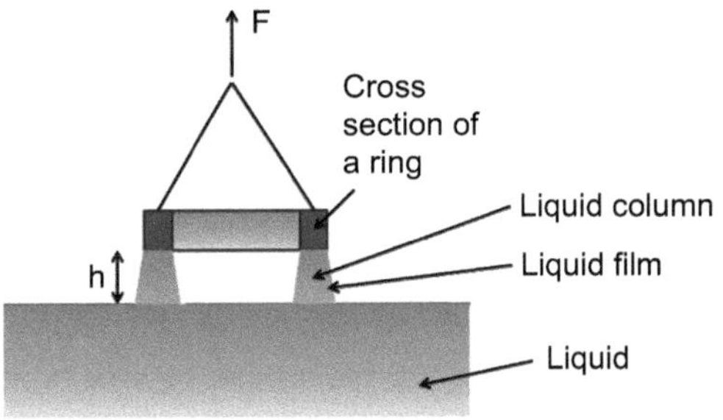

Figure 5.23 The du Noüy ring method for surface tension measurement. Reprinted with permission from Yasui et al. [9]. Copyright (2019), Elsevier.

5.4.2 Theoretical Study

According to the dynamic equilibrium model for a stable ultrafine bubble discussed previously, more than 50% of the surface of a stable ultrafine bubble is covered with a hydrophobic material such as oils and carbon particles [2]. Thus, it is expected that such ultrafine bubbles are preferentially trapped at the surface of liquid water (Fig. 5.24) [9]. One of the reasons is that a hydrophobic material repels water and that the free energy of a hydrophobic material is lower above the liquid surface than that underneath the liquid surface. The other reason is that an ultrafine bubble partly covered with a hydrophobic material hardly disappears at the liquid surface, while a normal gas bubble busts and disappears at the liquid surface.

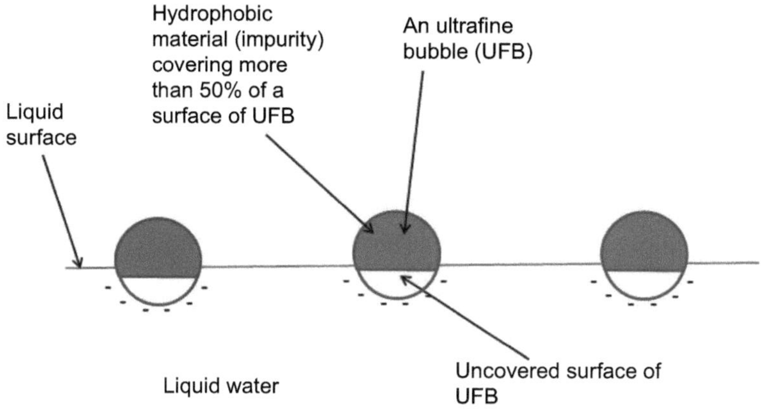

Figure 5.24 Ultrafine bubbles concentrated on a surface of liquid water. Each ultrafine bubble is partially covered with a hydrophobic material. Reprinted with permission from Yasui et al. [9]. Copyright (2019), Elsevier.

It has been experimentally reported that the hydrophobic polymeric particles of 113 nm in diameter considerably reduce the surface tension of liquid water when the area per unit particle mass at the liquid surface is less than 30 m^2/g [76]. The corresponding surface concentration of polymeric particles is in the order of 10^9 cm^{-2}. However, the surface concentration of ultrafine bubbles has been numerically estimated as less than 10^5 cm^{-2} when their bulk concentration has a typical value of 10^8 cm^{-3} [9]. Thus, the surface concentration of ultrafine bubbles is several orders of magnitude smaller than that required to reduce surface tension. Then, what is the reason for the decrease in surface tension by the presence of ultrafine bubbles measured by the du Noüy ring method? In the du Noüy ring method, surface tension is measured by the condition of the rupture of the liquid film under a ring (Fig. 5.23) [9, 77, 78]. When the uncovered surface of a stable ultrafine bubble is directed toward the gas above the liquid surface, the bubble bursts and disappears like a normal gas bubble. At the same time, the liquid film is ruptured. Thus, it is expected that the value of surface tension measured by the du Noüy ring method decreases by the presence of ultrafine bubbles because the rupture of the liquid film is accelerated [9]. It may be also related to the increase in the washing rate of a soiled cloth by the presence of ultrafine bubbles discussed in the previous subsection. In order to wash a cloth, water should penetrate small holes between the fibers of a cloth. However, due to surface tension, it is sometimes difficult. If the rupture of liquid film is accelerated by the presence of ultrafine bubbles, water more easily penetrates the small holes between cloth fibers, and a cloth is more effectively washed. Further studies are required on this topic.

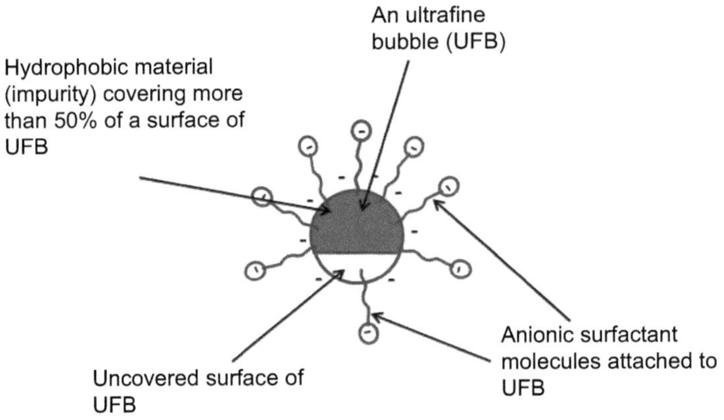

Figure 5.25 Anionic surfactant molecules attached to an ultrafine bubble negatively charged and partly covered with a hydrophobic material. Reprinted with permission from Yasui et al. [9]. Copyright (2019), Elsevier.

Next, the influence of a surfactant is discussed on the effect of ultrafine bubbles on "surface tension." It has been experimentally reported that an ultrafine bubble is negatively charged with a zeta potential of −40 mV [18–20]. The surface of a hydrophobic material partly covering the bubble surface can be also negatively charged due to the presence of a small amount of fatty acid in oils and fats, or some functional groups on a surface of carbon particles such as –COOH, –OH, and –CHO [79–82]. Thus, it is expected that the anionic part of an anionic surfactant is repelled from an ultrafine bubble as shown in Fig. 5.25 [9]. Although the collision frequency of an ultrafine bubble covered with a surfactant to the liquid surface is decreased compared to an ultrafine bubble without any surfactant, the trapping probability of an ultrafine bubble to the liquid surface is almost unchanged because the local hydrophobicity of a hydrophobic material covering an ultrafine bubble is almost unchanged. As a result, ultrafine bubbles are expected to be still concentrated at a liquid surface due to the hydrophobic part. This may be the reason for the decrease of about 20% in surface tension measured by the du Noüy ring method in an anionic surfactant solution by the presence of ultrafine bubbles because a liquid film is more easily ruptured by the presence of ultrafine bubbles at the liquid surface [9, 10].

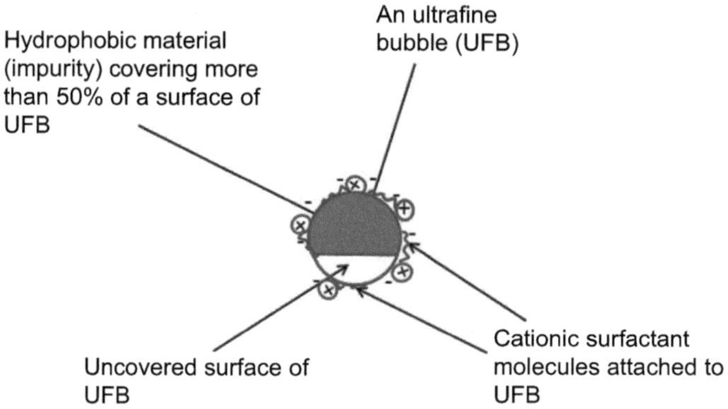

Figure 5.26 Cationic surfactant molecules attached to an ultrafine bubble negatively charged and partly covered with a hydrophobic material. Reprinted with permission from Yasui et al. [9]. Copyright (2019), Elsevier.

In a cationic surfactant solution, on the other hand, the cationic part of a surfactant is attracted to the negatively charged surface of an ultrafine bubble. As a result, the hydrophilic part of cationic surfactant is also attached to the hydrophobic material covering an ultrafine bubble (or to the uncovered part of an ultrafine bubble) (Fig. 5.26) [9]. In other words, cationic surfactant molecules adsorb in a horizontal orientation to an ultrafine bubble [83]. The local hydrophobicity of a hydrophobic material covering an ultrafine bubble is decreased because the number of hydrophilic groups on a hydrophobic material is almost doubled due to the attachment of cationic parts of surfactant molecules. In addition to the decrease in the collision frequency of an ultrafine bubble covered with a surfactant to the liquid surface compared to an ultrafine bubble without a surfactant, the trapping probability of an ultrafine bubble to the liquid surface decreases due to the decreased local hydrophobicity of a hydrophobic material covering an ultrafine bubble. It means that ultrafine bubbles are less concentrated at the liquid surface. It may be the reason why there is no effect of ultrafine bubbles on "surface tension" in an aqueous cationic surfactant solution [9, 10].

In conclusion, ultrafine bubbles partly covered with a hydrophobic material are concentrated at the surface of liquid water [9]. They do not burst and disappear at the liquid surface because more than 50% of the surface of an ultrafine bubble is covered with a hydrophobic material according to the dynamic equilibrium model for a stable ultrafine bubble (Fig. 5.24) [2]. Surface tension is unchanged by the presence of ultrafine bubbles under the typical concentration of ultrafine bubbles (less than about 10^{10}/mL). However, the rupture of a liquid film could be accelerated by the presence of ultrafine bubbles at the liquid surface because an ultrafine bubble bursts and disappears when the uncovered surface of an ultrafine bubble is directed toward the gas side of the liquid surface. It reduces the value of "surface tension" measured by the du Noüy ring method [9]. In a cationic surfactant solution, ultrafine bubbles are less concentrated on the liquid surface, and there is almost no effect of ultrafine bubbles on "surface tension." In an anionic surfactant solution, on the other hand, ultrafine bubbles are still concentrated on the liquid surface, and the rupture of a liquid film is still accelerated by the presence of ultrafine bubbles. The accelerated rupture of a liquid film may also result in the increase in the washing rate of a soiled cloth by the presence of ultrafine bubbles because liquid water more easily penetrates the small holes between the fibers of a cloth.

5.5 Conclusion

According to the dynamic equilibrium model [2], an ultrafine bubble is stabilized against dissolution when more than 50% of its surface is covered with a hydrophobic material such as oils and carbon particles. It has been partly confirmed by the TEM observation of ultrafine bubbles in aqueous solution without freezing [3]. With regard to OH radical production from ultrafine bubbles after stopping cavitation for the generation of micro- and ultrafine bubbles, the chemical reaction of H_2O_2 and O_3 in the liquid produced from violently collapsed bubbles during cavitation may result in the production of OH radicals after stopping cavitation [4–6]. Especially, in acidic conditions, the reaction of H_2O_2 and O_3 in the liquid is relatively slow, and OH radical production considerably continues even after stopping cavitation [4, 5, 74, 75]. With regard to the effect of ultrafine bubbles on the surface tension of liquid water, there should be no change in surface tension under the typical concentration of ultrafine bubbles (less than 10^{10}/mL) [9]. Instead, the rupture of a liquid film could be accelerated by the presence of ultrafine bubbles on the liquid surface when an ultrafine bubble bursts by directing the uncovered surface of an ultrafine bubble toward the gas side of the liquid surface [9]. It reduces the value of "surface tension" measured by the du Noüy ring method [9, 10]. The acceleration of the rupture of a liquid film may increase the washing rate of a soiled cloth by accelerating the penetration of liquid water into the small holes between the fibers of a cloth.

Acknowledgments

The author would like to thank Wataru Kanematsu, Toru Tuziuti, Nobuhiro Aya, Noriya Izu, and Kazumi Kato for useful discussions. A part of the study was supported by the Ministry of Economy, Trade and Industry (METI), Japan, as part of the international standardization project for fine bubbles.

References

1. Yasui, K., Tuziuti, T., and Kanematsu, W. (2018). Mysteries of bulk nanobubbles (ultrafine bubbles); stability and radical formation, *Ultrason. Sonochem.*, **48**, pp. 259–266.
2. Yasui, K., Tuziuti, T., Kanematsu, W., and Kato, K. (2016). Dynamic equilibrium model for a bulk nanobubble and a microbubble partly covered with hydrophobic material, *Langmuir*, **32**, pp. 11101–11110.
3. Sugano, K., Miyoshi, Y., and Inazato, S. (2017). Study of ultrafine bubble stabilization by organic material adhesion, *Jpn. J. Multiphase Flow*, **31**, pp. 299–306.

4. Yasui, K., Tuziuti, T., and Kanematsu, W. (2019). Mechanism of OH radical production from ozone bubbles in water after stopping cavitation, *Ultrason. Sonochem.*, **58**, 104707.
5. Takahashi, M., Chiba, K., and Li, P. (2007). Free-radical generation from collapsing microbubbles in the absence of a dynamic stimulus, *J. Phys. Chem. B*, **111**, pp. 1343–1347.
6. Liu, S., Oshita, S., Kawabata, S., Makino, Y., and Yoshimoto, T. (2016). Identification of ROS produced by nanobubbles and their positive and negative effects on vegetable seed germination, *Langmuir*, **32**, pp. 11295–11302.
7. Yasui, K., Tuziuti, T., and Kanematsu, W. (2016). Extreme conditions in a dissolving air nanobubble, *Phys. Rev. E*, **94**, 013106.
8. Yasui, K., Tuziuti, T., and Kanematsu, W. (2019). High temperature and pressure inside a dissolving oxygen nanobubble, *Ultrason. Sonochem.*, **55**, pp. 308–312.
9. Yasui, K., Tuziuti, T., Izu, N., and Kanematsu, W. (2019). Is surface tension reduced by nanobubbles (ultrafine bubbles) generated by cavitation? *Ultrason. Sonochem.*, **52**, pp. 13–18.
10. Ushida, A., Hasegawa, T., Takahashi, N., Nakajima, T., Murao, S., Narumi, T., and Uchiyama, H. (2012). Effect of mixed nanobubble and microbubble liquids on the washing rate of cloth in an alternating flow, *J. Surfact. Deterg.*, **15**, pp. 695–702.
11. Alheshibri, M., Qian, J., Jehannin, M., and Craig, V. S. J. (2016). A history of nanobubbles, *Langmuir*, **32**, pp. 11086–11100.
12. Yasui, K. (2018). *Acoustic Cavitation and Bubble Dynamics* (Springer, Switzerland).
13. Yasui, K. (2016). Mechanism for stability of ultrafine bubbles, *Jpn. J. Multiphase Flow*, **30**, pp. 19–26 (in Japanese).
14. Yasui, K. (2015). Dynamics of acoustic bubbles, Chapter 3, in *Sonochemistry and the Acoustic Bubble*, eds. Grieser, F., Choi, P. K., Enomoto, N., Harada, H., Okitsu, K., and Yasui, K. (Elsevier, Amsterdam), pp. 41–83.
15. Alheshibri, M. and Craig, V. S. J. (2018). Differentiating between nanoparticles and nanobubbles by evaluation of the compressibility and density of nanoparticles, *J. Phys. Chem. C*, **122**, pp. 21998–22007.
16. Alheshibri, M. and Craig, V. S. J. (2019). Generation of nanoparticles upon mixing ethanol and water; nanobubbles or not? *J. Colloid Interface Sci.*, **542**, pp. 136–143.
17. Takahashi, M. (2005). ζ potential of microbubbles in aqueous solutions: Electrical properties of the gas–water interface, *J. Phys. Chem. B*, **109**, pp. 21858–21864.

18. Azevedo, A., Etchepare, R. Calgaroto, S., and Rubio, J. (2016). Aqueous dispersion of nanobubbles: Generation, properties and features, *Minerals Engineering*, **94**, pp. 29–37.

19. Cho, S. H., Kim, J. Y., Chun, J. H., and Kim, J. D. (2005). Ultrasonic formation of nanobubbles and their zeta-potentials in aqueous electrolyte and surfactant solutions, *Colloid Surf. A*, **269**, pp. 28–34.

20. Kim, J. Y., Song, M. G., and Kim, J. D. (2000). Zeta potential of nanobubbles generated by ultrasonication in aqueous alkyl polyglycoside solutions, *J. Colloid Interface Sci.*, **223**, pp. 285–291.

21. Temperley, H. N. V. (1947). The behaviour of water under hydrostatic tension: III, *Proc. Phys. Soc.*, **59**, pp. 199–208.

22. Fisher, J. C. (1948). The fracture of liquids, *J. Appl. Phys.*, **19**, pp. 1062–1067.

23. Galloway, W. J. (1954). An experimental study of acoustically induced cavitation in liquids, *J. Acoust. Soc. Am.*, **26**, pp. 849–857.

24. Leighton, T. G. (1994). *The Acoustic Bubble* (Academic Press, London).

25. Young, F. R. (1999). *Cavitation* (Imperial College Press, London).

26. Fox, F. E. and Herzfeld, K. F. (1954). Gas bubbles with organic skin as cavitation nuclei, *J. Acoust. Soc. Am.*, **26**, pp. 984–989.

27. Yount, D. E. (1979). Skins of varying permeability: A stabilization mechanism for gas cavitation nuclei, *J. Acoust. Soc. Am.*, **65**, pp. 1429–1439.

28. Yount, D. E., Gillary, E. W., and Hoffman, D. C. (1984). A microscopic investigation of bubble formation nuclei, *J. Acoust. Soc. Am.*, **76**, pp. 1511–1521.

29. Apfel, R. E. (1970). The role of impurities in cavitation-threshold determination, *J. Acoust. Soc. Am.*, 48, pp. 1179–1186.

30. Crum, L. A. (1979). Tensile strength of water, *Nature (London)*, **278**, pp. 148–149.

31. Atchley, A. A. and Prosperetti, A. (1989). The crevice model of bubble nucleation, *J. Acoust. Soc. Am.*, **86**, pp. 1065–1084.

32. Madanshetty, S. I. and Apfel, R. E. (1991). Acoustic microcavitation: Enhancement and applications, *J. Acoust. Soc. Am.*, **90**, pp. 1508–1514.

33. Tuziuti, T., Yasui, K., Sivakumar, M., Iida, Y., and Miyoshi, N. (2005). Correlation between acoustic cavitation noise and yield enhancement of sonochemical reaction by particle addition, *J. Phys. Chem. A*, **109**, pp. 4869–4872.

34. Mohamedi, G., Azmin, M., Pastoriza-Santos, I., Huang, V., Perez-Juste, J., Liz-Marzan, L. M., Edirisinghe, M., and Stride, E. (2012). Effects of

gold nanoparticles on the stability of microbubbles, *Langmuir*, **28**, pp. 13808–13815.

35. Azmin, M., Mohamedi, G., Edirisinghe, M., and Stride, E. P. (2012). Dissolution of coated microbubbles: The effect of nanoparticles and surfactant concentration, *Mater. Sci. Eng. C*, **32**, pp. 2654–2658.
36. Weijs, J. H., Seddon, J. R. T., and Lohse, D. (2012). Diffusive shielding stabilizes bulk nanobubble clusters, *ChemPhysChem*, **13**, pp. 2197–2204.
37. Steitz, R., Gutberlet, T., Hauss, T., Klosgen, B., Krastev, R., Schemmel, S., Simonsen, A. C., and Findenegg, G. H. (2003). Nanobubbles and their precursor layer at the interface of water against a hydrophobic substrate, *Langmuir*, **19**, pp. 2409–2418.
38. Mezger, M., Schoder, S., Reichert, H., Schroder, H., Okasinski, J., Honkimaki, V., Ralston, J., Bilgram, J., Roth, R., and Dosch, H. (2008). Water and ice in contact with octadecyl-trichlorosilane functionalized surfaces: A high resolution x-ray reflectivity study, *J. Chem. Phys.*, **128**, 244705.
39. Lu, Y. H., Yang, C. W., and Hwang, I. S. (2012). Molecular layer of gas-like domains at a hydrophobic-water interface observed by frequency-modulation atomic force microscopy, *Langmuir*, **28**, pp. 12691–12695.
40. Peng, H., Birkett, G. R., and Nguyen, A. V. (2013). Origin of interfacial nanoscopic gaseous domains and formation of dense gas layer at hydrophobic solid-water interface, *Langmuir*, **29**, pp. 15266–15274.
41. Azadi, M., Nguyen, A. V., and Yakubov, G. E. (2015). Attractive forces between hydrophobic solid surfaces measured by AFM on the first approach in salt solutions and in the presence of dissolved gases, *Langmuir*, **31**, pp. 1941–1949.
42. Peng, H., Hampton, M. A., and Nguyen, A. V. (2013). Nanobubbles do not sit alone at the solid-liquid interface, *Langmuir*, **29**, pp. 6123–6130.
43. Dammer, S. M. and Lohse, D. (2006). Gas enrichment at liquid-wall interfaces, *Phys. Rev. Lett.*, **96**, 206101.
44. Petsev, N. D., Shell, M. S., and Leal, L. G. (2013). Dynamic equilibrium explanation for nanobubbles' unusual temperature and saturation dependence, *Phys. Rev. E*, **88**, 010402 (R).
45. Yasui, K., Tuziuti, T., Kanematsu, W., and Kato, K. (2015). Advanced dynamic-equilibrium model for a nanobubble and a micropancake on a hydrophobic or hydrophilic surface, *Phys. Rev. E*, **91**, 033008.
46. Israelachvili, J. N. (2011). *Intermolecular and Surface Forces*, 3rd ed. (Elsevier, Amsterdam).

47. Chan, C. U. and Ohl, C. D. (2012). Total-internal-reflection-fluorescence microscopy for the study of nanobubble dynamics, *Phys. Rev. Lett.*, **109**, 174501.

48. Dietrich, E., Zandvliet, H. J. W., Lohse, D., and Seddon, J. R. T. (2013). Particle tracking around surface nanobubbles, *J. Phys.: Condens. Matter*, **25**, 184009.

49. Lohse, D. and Zhang, X. (2015). Surface nanobubbles and nanodroplets, *Rev. Mod. Phys.*, **87**, pp. 981–1035.

50. Brenner, M. P. and Lohse, D. (2008). Dynamic equilibrium mechanism for surface nanobubble stabilization, *Phys. Rev. Lett.*, **101**, 214505.

51. Kondepudi, D. and Prigogine, I. (1998). *Modern Thermodynamics* (John Wiley & Sons, Chichester, U.K.).

52. Modell, M. and Reid, R. C. (1974). *Thermodynamics and Its Applications* (Prentice Hall, Englewood Cliffs, NJ).

53. Tada, K., Maeda, M., Nishiuchi, Y., Nagahara, J., Hata, T., Zhuowei, Z., Yoshida, Y., Watanabe, S., and Ohmori, M. (2014). ESR measurement of hydroxyl radicals in micro-nanobubble water, *Chem. Lett.*, **43**, pp. 1907–1908.

54. Makino, K., Mossoba, M. M., and Riesz, P. (1982). Chemical effects of ultrasound on aqueous solutions: Evidence for OH and H by spin trapping, *J. Am. Chem. Soc.*, **104**, pp. 3537–3539.

55. Brenner, M. P., Hilgenfeldt, S., and Lohse, D. (2002). Single-bubble sonoluminescence, *Rev. Mod. Phys.*, **74**, pp. 425–484.

56. Henglein, A. (1993). Contributions to various aspects of cavitation chemistry, in *Advances in Sonochemistry* vol. 3, ed. Mason, T. J. (JAI Press, London), pp. 17–83.

57. Krishnan, J. S., Dwivedi, P., and Moholkar, V. S. (2006). Numerical investigation into the chemistry induced by hydrodynamic cavitation, *Ind. Eng. Chem.*, **45**, pp. 1493–1504.

58. Yasui, K. and Kato, K. (2012). Bubble dynamics and sonoluminescence from helium or xenon in mercury and water, *Phys. Rev. E*, **86**, 036320. 86, 069901 (E).

59. Gogate, P. R., Shirgaonkar, I. Z., Sivakumar, M., Senthilkumar, P., Vichare, N. P., and Pandit, A. B. (2001). Cavitation reactors: Efficiency assessment using a model reaction, *AIChE J.*, **47**, pp. 2526–2538.

60. Koda, S., Kimura, T., Kondo, T., and Mitome, H. (2003). A standard method to calibrate sonochemical efficiency of an individual reaction system, *Ultrason. Sonochem.*, **10**, pp. 149–156.

61. Yasui, K. (2016). Unsolved problems in acoustic cavitation, in *Handbook of Ultrasonics and Sonochemistry* vol. 1, ed. Ashokkumar, M. (Springer, Singapore), pp. 259–292.

62. Yasui, K., Tuziuti, T., Sivakumar, M., and Iida, Y. (2004). Sonoluminescence, *Appl. Spectrosc. Rev.*, **39**, pp. 399–436.

63. Barber, B. P., Hiller, R. A., Lofstedt, R., Putterman, S. J., and Weninger, K. R. (1997). Defining the unknowns of sonoluminescence, *Phys. Rep.*, **281**, pp. 65–143.

64. Young, F. R. (2005). *Sonoluminescence* (CRC Press, Boca Raton, USA).

65. Gaitan, D. F., Crum, L. A., Church, C. C., and Roy, R. A. (1992). Sonoluminescence and bubble dynamics for a single, stable, cavitation bubble, *J. Acoust. Soc. Am.*, **91**, pp. 3166–3183.

66. Yoshioka, K. and Omura, A. (1962). The light emission from a single bubble driven by ultrasound and the spectra of acoustic oscillations, *Proc. Annu. Meet. Acoust. Soc. Jpn.*, pp. 125–126 (in Japanese).

67. Didenko, Y. T. and Suslick, K. S. (2002). The energy efficiency of formation of photons, radicals and ions during single-bubble cavitation, *Nature (London)*, **418**, pp. 394–397.

68. Yasui, K., Tuziuti, T., Sivakumar, M., and Iida, Y. (2005). Theoretical study of single-bubble sonochemistry, *J. Chem. Phys.*, **122**, 224706.

69. Yasui, K. (1997). Alternative model of single-bubble sonoluminescence, *Phys. Rev. E*, **56**, pp. 6750–6760.

70. Yasui, K. (1996). Variation of liquid temperature at bubble wall near the sonoluminescence threshold, *J. Phys. Soc. Jpn.*, **65**, pp. 2830–2840.

71. Yasui, K. (2001). Effect of liquid temperature on sonoluminescence, *Phys. Rev. E*, **64**, 016310.

72. Lohse, D., Brenner, M. P., Dupont, T. F., Hilgenfeldt, S., and Johnston, B. (1997). Sonoluminescing air bubbles rectify argon, *Phys. Rev. Lett.*, **78**, pp. 1359–1362.

73. Matula, T. J. and Crum, L. A. (1998). Evidence of gas exchange in single-bubble sonoluminescence, *Phys. Rev. Lett.*, **80**, pp. 865–868.

74. Staehelin, J. and Holgne, J. (1982). Decomposition of ozone in water: Rate of initiation by hydroxide ions and hydrogen peroxide, *Environ. Sci. Technol.*, **16**, pp. 676–681.

75. Taube, H. and Bray, W. C. (1940). Chain reactions in aqueous solutions containing ozone, hydrogen peroxide and acid, *J. Am. Chem. Soc.*, **62**, pp. 3357–3373.

76. Wolert, E., Setz, S. M., Underhill, R. S., Duran, R. S., Schappacher, M., Deffieux, A., Holderle, M., and Mulhaupt, R. (2001). Meso- and microscopic behavior of spherical polymer particles assembling at the air-water interface, *Langmuir*, **17**, pp. 5671–5677.

77. Adamson, A. W. (1990). *Physical Chemistry of Surfaces*, 5th ed. (John Wiley & Sons, New York).

78. Butt, H. J., Graf, K., and Kappl, M. (2013). *Physical and Chemistry of Interfaces*, 3rd ed. (Wiley-VCH, Germany).

79. Williams, H. J. (1982). *Introduction to Organic Chemistry* (John Wiley & Sons, Chichester, U.K.).

80. Boehm, H. P. (1966). Chemical identification of surface groups, *Adv. Catal.*, **16**, pp. 179–274.

81. Donnet, J. B. (1968). The chemical reactivity of carbons, *Carbon*, **6**, pp. 161–176.

82. Puri, B. R. (1970). Surface complexes on carbons, in *Chemistry and Physics of Carbon* vol. 6, ed. Walker, P. L. (Marcel Dekker, New York), pp. 191–282.

83. Hunter, R. J. (1981). *Zeta Potential in Colloid Science, Principles and Applications* (Academic Press, London).

Chapter 6

Study of Ultrafine Bubble Stabilization by Organic Material Adhesion

Kou Sugano, Yuichi Miyoshi, and Sachiko Inazato
Panasonic Corporation, 1006 Kadoma, Kadoma City, Osaka 571-8506, Japan
sugano.k@jp.panasonic.com

We confirmed the effect of organic material adhesion as ultrafine bubble (UFB) stabilization mechanism. An organic material was added to generate UFBs, and the state of UFB particles was investigated by the dynamic light scattering method, transmission electron microscopy (TEM), and resonance mass measurement method. Disappearance of UFBs was suppressed by the addition of organic material, and it was confirmed that the organic material adhered to the surface of UFBs. We concluded that the adhesion of organic material inhibited gas dissolution from bubbles.

Ultrafine Bubbles
Edited by Koichi Terasaka, Kyuichi Yasui, Wataru Kanematsu, and Nobuhiro Aya

ISBN 978-981-4877-59-6 (Hardcover), 978-1-003-14195-2 (eBook)
www.jennystanford.com

6.1 Introduction

Fine bubbles can be generated by cavitation using a Venturi tube or swirling flow, and water becomes opaque due to the generation of many micro- and milli-bubbles. When the opaque liquid is left over with the microbubble generator off, the bubble rises to the surface level due to buoyancy or contracts and disappears in the water and the liquid returns to transparent water. When this transparent water is measured by the dynamic light scattering method, the existence of UFB having a diameter of about 100 to 200 nm may be confirmed. It is reported that this UFB exists stably for more than 1 month [1]. However, the mechanism by which a UFB stabilizes in water remains unknown, and several theories have been proposed [1–10], but a clear mechanism has not yet been established in reality.

In the dynamic light scattering method, it is difficult to clarify the mechanism since bubbles and solid particles cannot be distinguished from each other. In order to promote the elucidation of the mechanism, a device configuration with high cleanliness is required as much as possible. Various liquid impurities may be involved in the occurrence of UFBs. The authors developed a UFB generator in a clean environment excluding solid particles and metal ions and tried to elucidate the stabilization mechanism of UFBs in order to promote the industrial use of UFBs. As a result, we report here the findings on the UFB stabilization mechanism, which suppresses the annihilation of bubbles.

6.2 Model of Stabilization Mechanism

The authors confirmed the phenomenon that UFB concentration gradually decreased in the clean generator. Details are provided in Section 6.4. It was also found that the generated UFB was stable without a change in concentration even after several days. Since the UFB concentration gradually decreases, some substance is eluted from the wetted member, which acts on the bubble. The UFB concentration also decreases as the elution amount decreases. The model of the UFB stabilization mechanism is shown in Fig. 6.1.

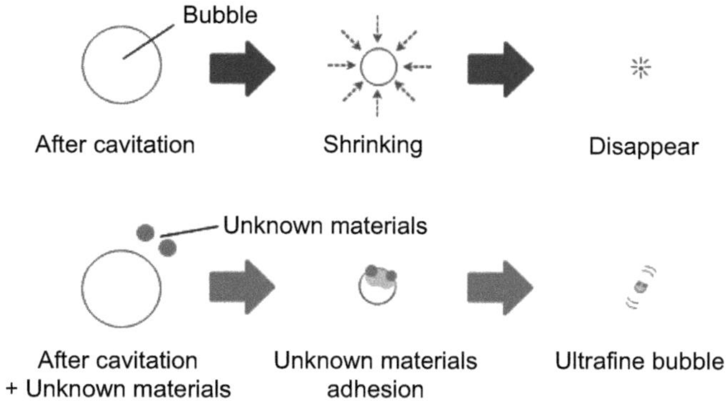

Figure 6.1 UFB stabilization model.

Normally, bubbles generated by cavitation contract and disappear (Fig. 6.1, upper row), but if any eluted material is present, it adheres to the bubble. Although the bubble shrinks, the adhered material covers the entire bubble and stabilizes with a particle size of several hundred nanometers or less in diameter (Fig. 6.1, lower row). The bubble takes the route of shrinkage, that is, the gas within the bubble elutes into the surrounding liquid. Adhered substances are thought to inhibit this route. This model includes not only the dynamic equilibrium model [1] but also the feature of the skin model [2] that the whole bubble is covered.

6.3 Experimental Method

The configuration of the UFB generator used in the experiment for verifying the UFB stabilization model is shown in Fig. 6.2. Further, a list of components is shown in Table 6.1.

The generator was installed in a class 10 clean booth, and UPW (ultrapure water, specific resistance value 18 MΩ · cm) was used as the water to be supplied. The dissolved oxygen concentration of UPW is 4~5 mg/L (experimental environment measurement value), and UFB water produced in one experiment is 60 ml. UPW sealed in the generator was circulated by a pump and passed through a cavitation nozzle to generate UFB.

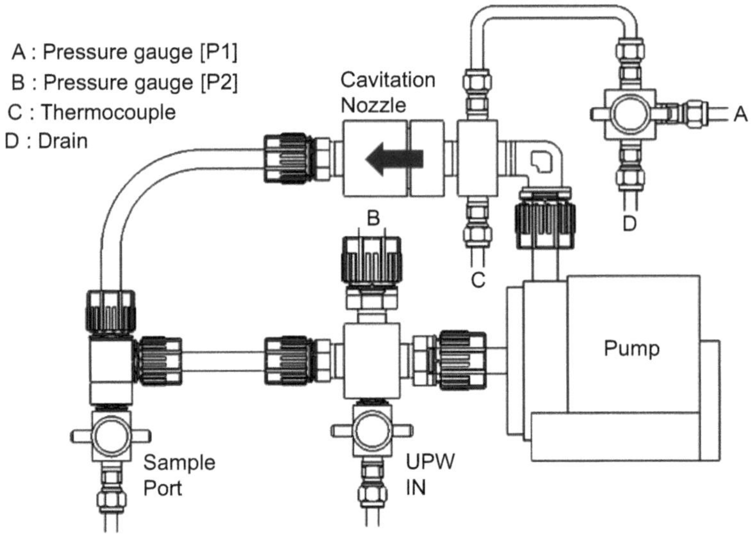

Figure 6.2 Construction of a UFB generator.

Table 6.1 List of components

Component parts	Material in contact	Manufacturing
Pump	PTFE, PFA	LEVITRONIX
Tube	PFA	Yodogawa Hu-Tech
Tube joint	PFA	PILLAR
O-ring	EPDM	–
Thermocouple	PFA	N D S
Pressure gauge	PTFE, PFA	SURPASS
Adapter	PTFE	Custom made
Nozzle	PTFE	Custom made

PTFE = PolyTetraFluoroEthylene
PFA = PerFluoroAlkoxy (vinyl ether)
EPDM = Ethylene-Propylene-Diene ternary copolymer

In order to prevent contamination, the UFB generator adopts a method of generating UFB from dissolved gas without providing an open water tank or a mechanism for sucking gas. Therefore, it is not UFB generation in air supersaturated state. The impeller of the pump was magnetically levitated and had no dust generation, and the part to be wetted was made of fluorocarbon resin material except for the ethylene propylene diene monomer (EPDM)-made O-ring.

The pressure value was measured by installing a pressure gauge at the point immediately before the cavitation nozzle [P1] and immediately before the pump [P2]. The UFB concentration was measured after releasing the pressure inside the generator using a Nano Sight LM-10 manufactured by Malvern Co., Ltd., which is a method of collecting UFB water from the Sample Port and analyzing the nano tracking analysis method.

The pressure inside the generator was released beforehand so that the influence on gas dissolution due to pressure change was made as small as possible. Measurement was made 15 min after the pump driving time at which concentration change stabilized. The temperature of the liquid was measured by installing a thermocouple just before the cavitation nozzle; the initial values of 25~27°C changed only to 37~40°C after 15 min (Δt = 12~13°C). Table 6.2 shows the conditions at the time of UFB generation.

Table 6.2 Condition of generation

Generation time	15	min
Pump rotation speed	8000	rpm
Liquid pressure [P1]		
Initial pressure	0	kPa[G]
Exit pressure	180~200	kPa[G]
Liquid pressure [P2]		
Initial pressure	0	kPa[abs]
After 5 minutes	-15~-20	kPa[abs]
Exit pressure	30~37	kPa[abs]
Liquid temperature		
Initial temperature	25~27	degree C
Exit temperature	37~40	degree C
Liquid flow	7.1~8.0	L/min

The cavitation nozzle shown in Fig. 6.3 is connected to a 3/8-inch PFA pipe with a joint, so it is inserted in the case and sealed with an O-ring. The nozzle body is a cylinder, with a diameter of 25.6 mm and a height of 20 mm.

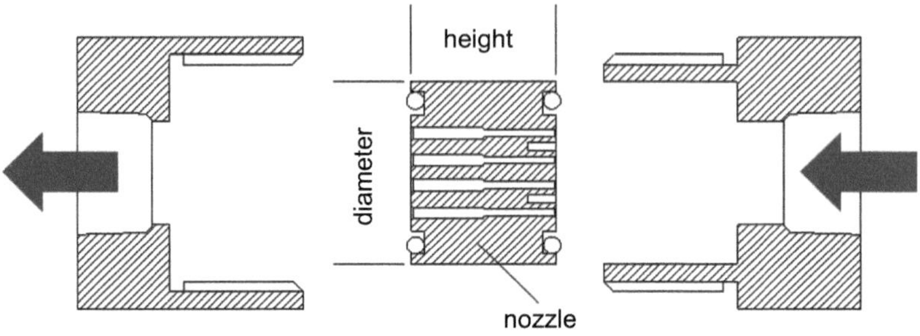

Figure 6.3 Cross section of a cavitation nozzle.

Pressurized by the pump, water passes through the PFA piping and is narrowed by the cavitation nozzle. The flow velocity increases, and the pressure decreases and returns to the PFA pipe diameter on the inlet side. Cavitation occurs, and bubbles are generated.

The definition formula of the Reynolds number *Re* is as follows:

$$Re = \frac{VL}{v} \tag{6.1}$$

Here, V is the relative average speed of the flow, L is the characteristic length, and v is the kinematic viscosity coefficient. *Re* in this experimental system was 23,600.

The definition formula of the cavitation number σ is as follows:

$$\sigma = \frac{(2(P - Pv))}{dV^2} \tag{6.2}$$

where P is the absolute pressure, Pv is the vapor pressure, d is the density of the fluid, V is the representative velocity of the flow, and σ in this experimental system is 0.4. It is more effective for cavitation generation to increase L in order to increase the Reynolds number, but for general machines, the height of the nozzle body was set as 20 mm from the limit of processing technology.

Table 6.2 shows operating conditions such as pressure, temperature, and flow rate at the time of UFB generation. In this paper, UFB concentration was measured by Nano Sight LM -10.

6.4 Results

6.4.1 UFB Generation Number Dependency

Details of the UFB concentration lowering phenomenon mentioned in Section 6.2 are described as follows. A decrease in UFB concentration was confirmed with the repetition of UFB production in the apparatus. The UFB generation procedure is shown as follows, but it was found that the procedure ① to ⑤ was carried out as a set and was carried out eight times. The UFB concentration gradually decreased, as shown in Fig. 6.4.

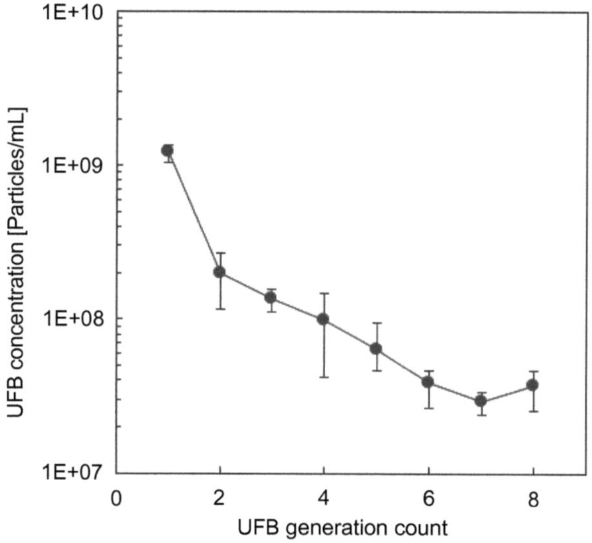

Figure 6.4 Transition of UFB concentration on generation count.

The nozzle diameter and the length of the hole of the nozzle were measured after the experiment, but the dimensions did not change. This is not the effect of nozzle structure change due to erosion.

- Experimental procedure for UFB generation
  ① UPW introduced
  ② Pump ON (Circulation for 15 min)
  ③ Pump OFF
  ④ UFB water sampling and measurement
  ⑤ Channel cleaning

Elution from fluorocarbon resin materials is well known in processes requiring care of foreign materials and metals like in the semiconductor industry. It is common to use acid cleaning beforehand to wash out the elute [11]. In this experiment, UFB was generated without cleaning the fluorocarbon resin with acid, so it is considered that some eluted material is present from the wetted member and affects UFB generation. In the aforementioned experiments, the amount of elution decreased, and as a result it was considered that the UFB concentration decreased.

In order to confirm this, the eluate of UFB water obtained in the experiment was analyzed. The UFB water shown in Fig. 6.4 was analyzed for inorganic and organic material by inductively coupled plasma-mass spectrometry (ICP-MS), ion chromatography (IC), and total organic carbon (TOC). The highest concentration UFB water in Fig. 6.4 was analyzed. UPW was analyzed as this reference.

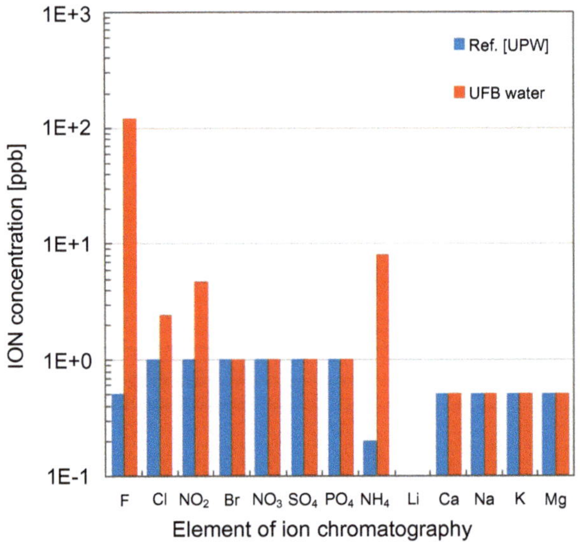

Figure 6.5 Analysis result by ion chromatography.

In the analysis of inorganic substances, F, Cl, NO_2, and NH_4 ions were detected by IC, and Zn was detected by ICP-MS in the order of ppb. Figure 6.5 shows the detection result by IC, and Fig. 6.6 shows the detection result by ICP-MS. F, NH_4, and Zn were added to 1 ppm UPW each to generate UFB.

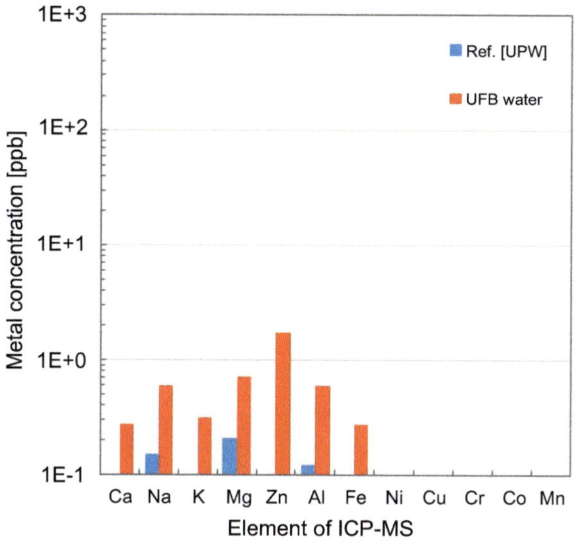

Figure 6.6 Analysis result by ICP-MS.

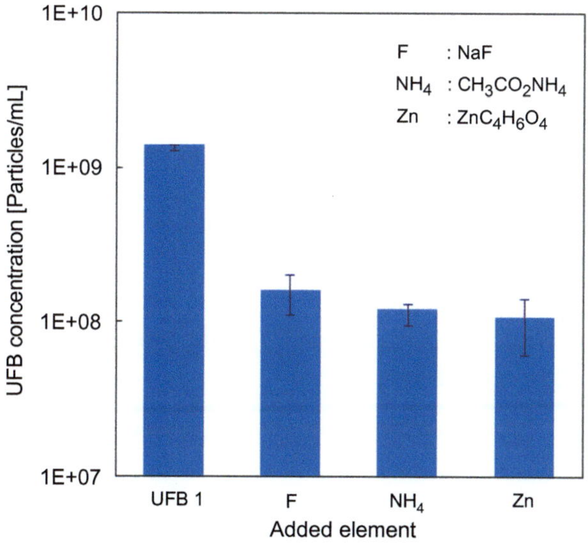

Figure 6.7 Influence of each metal element addition on UFB concentration.

As a result, as shown in Fig. 6.7, no element contributed to the UFB concentration.

The TOC was then measured to analyze the organics. As a result, it was found that the TOC concentration was correlated with the UFB concentration.

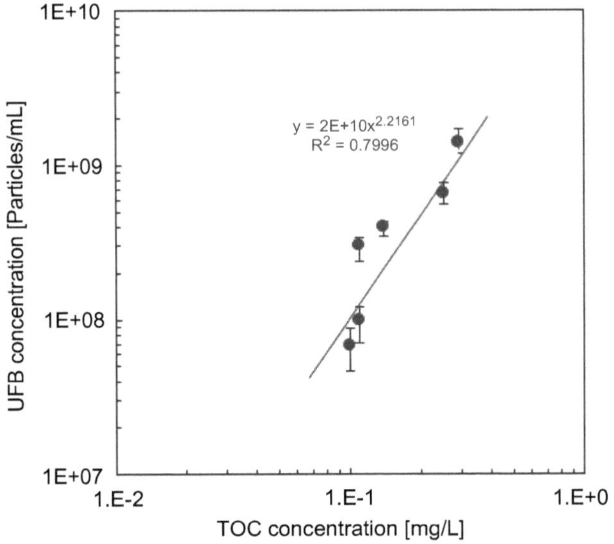

Figure 6.8 Relationship between TOC concentration and UFB concentration.

Figure 6.8 shows the results of TOC analysis of UFB water with the number of experiments 1 to 6 in Fig. 6.4. Analysis of this organic substance itself was not carried out, but it showed that the organic material in the liquid acts on UFB and contributes to UFB concentration.

6.4.2 Effect of Addition of Organic Material on UFB Concentration

In order to verify the influence of TOC on UFB concentration, UFB was generated by adding organic material to UPW in advance. The organic material was quantitatively added from the drain port upstream of the cavitation nozzle. Table 6.3 shows the list of organic materials used in the experiment. Figure 6.9 shows the result of adding these organic materials and generating UFB. There was no dependence of TOC concentration on IPA and HMDS. We think this is due to the small molecular weight and short carbon chain. However,

UFB concentration tended to increase with heptane and dodecane depending on the added concentration, and data indicating that TOC in liquid contributed to suppression of UFB annihilation were obtained. It was also found that UFB concentration decreased as the addition exceeded a certain amount and peaked with respect to the added amount. Hydrocarbon has a surfactant effect and becomes self-micellar when it exceeds the critical micelle concentration, so the action on UFB decreases and the UFB concentration decreases.

Table 6.3 List of added organic materials

Organic material	Chemical formula	Note
IPA	C_3H_8O	2-propanol
HMDS	$C_6H_{18}OSi_2$	Bis(trimethylsily)amine
Heptane	C_7H_{16}	Linear alkane
Dodecane	$C_{12}H_{26}$	Linear alkane
Paraffin	C_nH_{2n+2} C > 20	Chain saturated hydrocarbon
Squalene	$C_{30}H_{50}$	2,6,10,15,19,23-hexamethyl-2,6,10,14,18,22-tetracosahexaene
Oleic Acid	$C_{18}H_{34}O_2$	ω–9 fatty acids

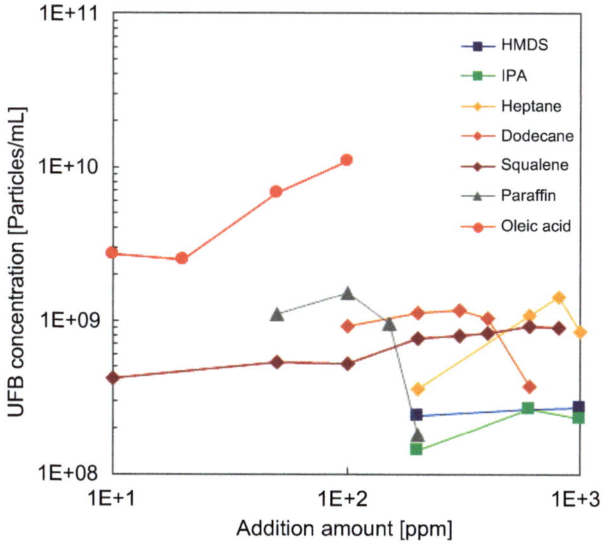

Figure 6.9 UFB concentration of the added amount of various organic materials.

Note that the self-micellarized hydrocarbons are considered to have a size of 50 nm or less in diameter, so they cannot be detected by the nano tracking analysis method.

From the results shown in Fig. 6.9, it has been found that when oleic acid is added, high concentrations of UFBs can be generated at low concentrations. From this result, a long-chain fatty acid, oleic acid, was added to generate UFBs. The results are shown in Fig. 6.10. The addition of 100 ppm resulted in a high concentration of 1.1×10^{10}/mL, indicating that the effect of suppressing UFB extinction was remarkable. The generated particle diameter did not differ even when the added amount became 10 times, and the peak particle diameter was around 80 nm in diameter. In the particle size distribution diagram shown in Fig. 6.11, the particle distribution range did not change, and no phenomenon of particle aggregation was observed. This graph is obtained by enlarging the 0 to 300 nm portion from the original graph upper right. Figure 6.12 shows the change in concentration when this UFB water is left standing at room temperature in a clean room, and Fig. 6.13 shows the particle size distributions. The evaluation was made for 76 days, and the particle size distributions were nearly in the same range as that after 0 day, which indicates that UFBs were nearly stabilized.

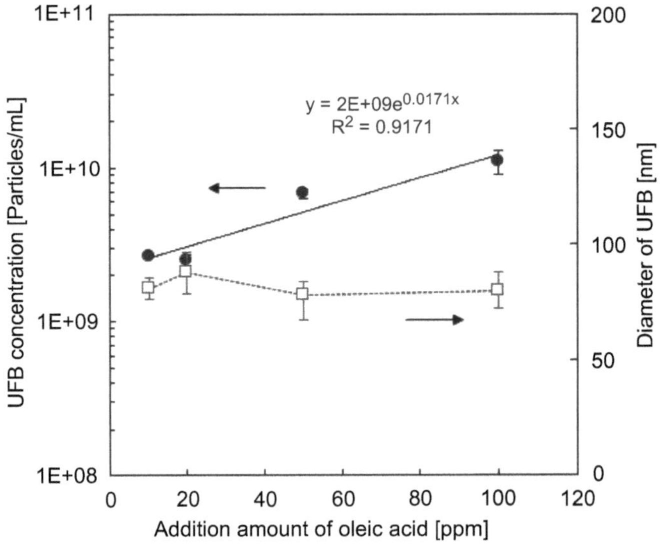

Figure 6.10 Dependence of UFB concentration and diameter on the added amount of oleic acid.

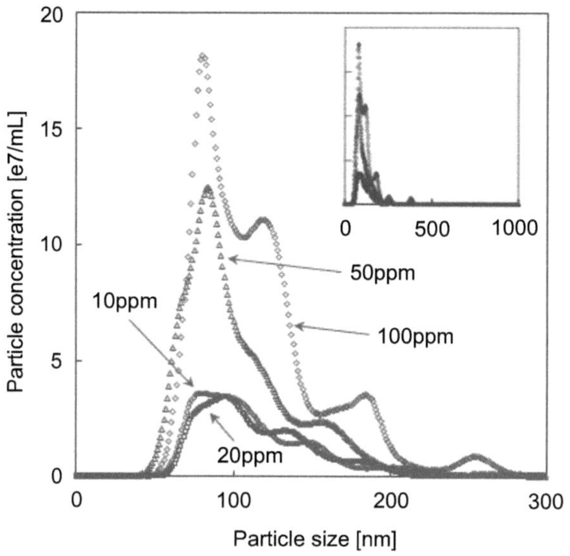

Figure 6.11 Particle size distribution of the added amount of oleic acid.

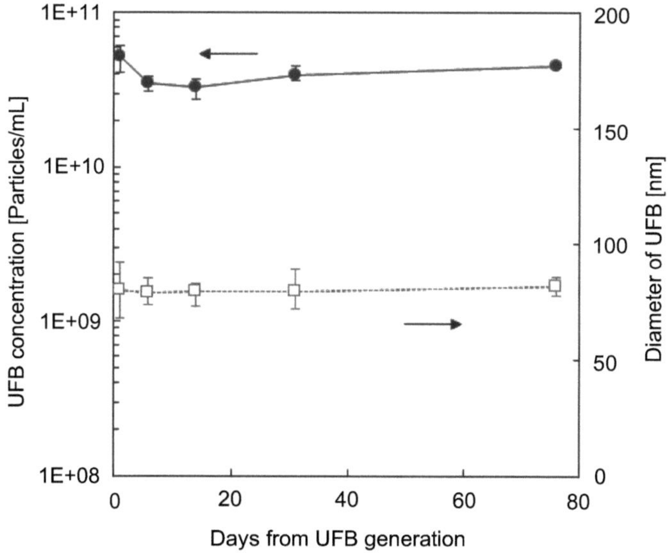

Figure 6.12 Time dependence of oleic acid added UFB concentration and diameter.

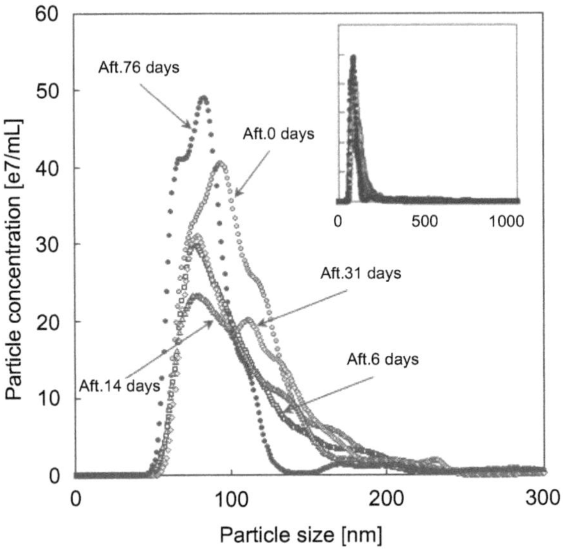

Figure 6.13 Particle size distribution over time of oleic acid added UFBs.

In addition, almost the same result was obtained when long-chain vitamin E (α-tocopherol) was added. The evaluation continued until the 433rd day.

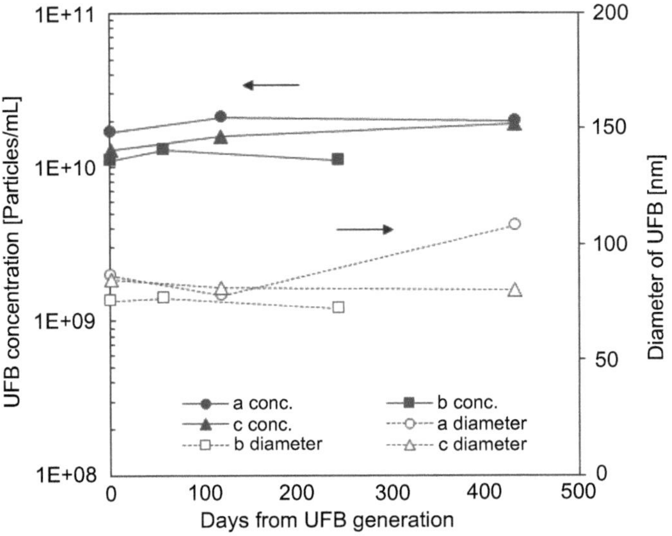

Figure 6.14 Time transition of various UFB concentration and diameter.

The change in concentration is shown in Fig. 6.14, and the particle size distribution is shown in Fig. 6.15.

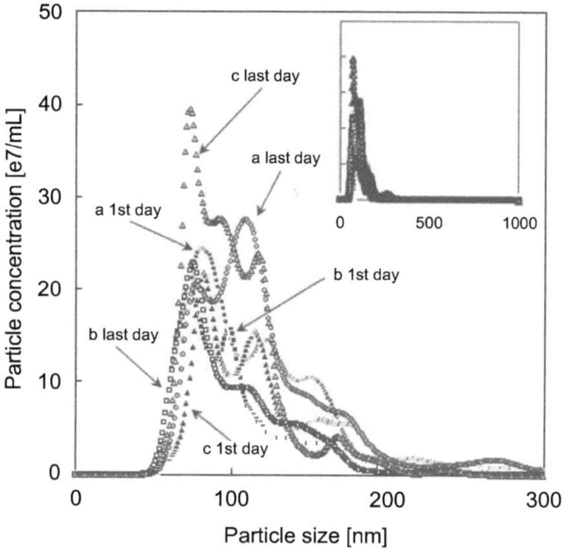

Figure 6.15 Particle size distribution over time of various UFBs.

Here, (a) is the result of addition of oleic acid in this experiment system, (b) is the result of addition of oleic acid in a pressure-dissolving generator that blows air into the open tank after intake of air and super saturation of the dissolved gas, and (c) is the result of addition of alpha tocopherol in this experimental system. The particle diameter of (a) seems to increase slightly, but the diameter is 108 nm and is maintained around 100 nm. From these results, it has been shown to remain stable for at least a year or more.

These samples were stored at locations affected by ambient temperature. It is considered from this experiment that bubbly extinction can be suppressed by adding organic material.

The influence of the hardness of water on the UFB concentration was also investigated. The results are shown in Fig. 6.16. The hardness was adjusted to 60 mg/L by adding Ca and Mg to UPW. Here, (A) is oleic acid added, (B) is adjusted water having a hardness of 60 mg/L, and (A + B) is addition of oleic acid to (B). Although (B) is equivalent to the UFB concentration of UPW, (A + B) has a single-digit UFB concentration lower than that of (a). From this result, it is considered that the hardness of water inhibits the bubble annihilation suppression function by the organic matter.

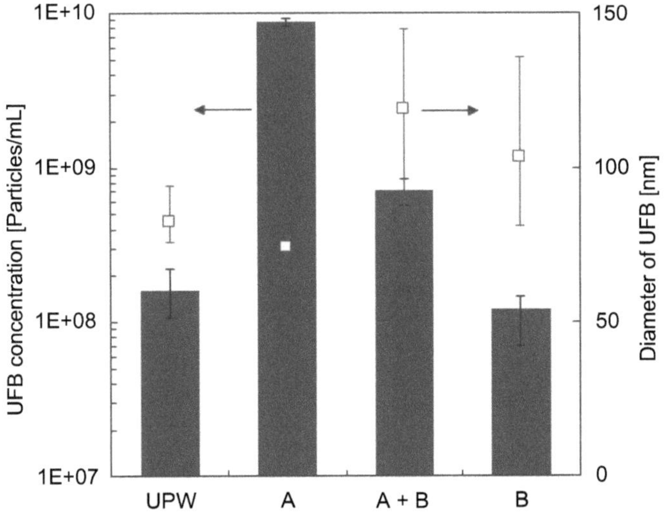

Figure 6.16 Influence of water hardness on UFB concentration.

6.4.3 UFB Analysis by TEM and Resonant Mass Measurement

We developed a technique to observe UFB directly with a transmission electron microscope (TEM) [12]. TEM observation was used for UFB generated using this technique. A microelectromechanical systems (MEMS) chip is mounted on an in situ holder of TEM. UFB water is introduced into the MEMS chip to make a very thin liquid phase of several hundred nanometers. By this, we observe the particles in the liquid with high resolution. This method is characterized by the fact that UFBs can be directly observed as they are without special treatment such as freezing. Photo 6.1 is a TEM image of UFB to which oleic acid is added, Photo 6.2 is a TEM image of UFB to which oleic acid and alkane are added, and Photo 6.3 is a TEM image of UFB to which α-tocopherol is added. In each case, adhesion of black material was confirmed on the surface of transparent round particles. This transparent particle is a bubble, and the black material is considered to be an organic material. It was possible to obtain a photograph suggesting that bubbles were stably formed by organic material adhering to the surface of bubbles.

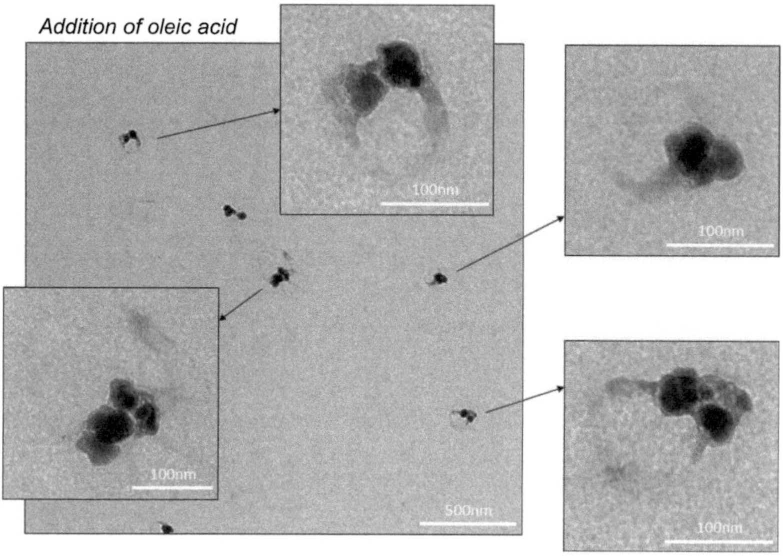

Photo 6.1 TEM image of oleic acid added UFB.

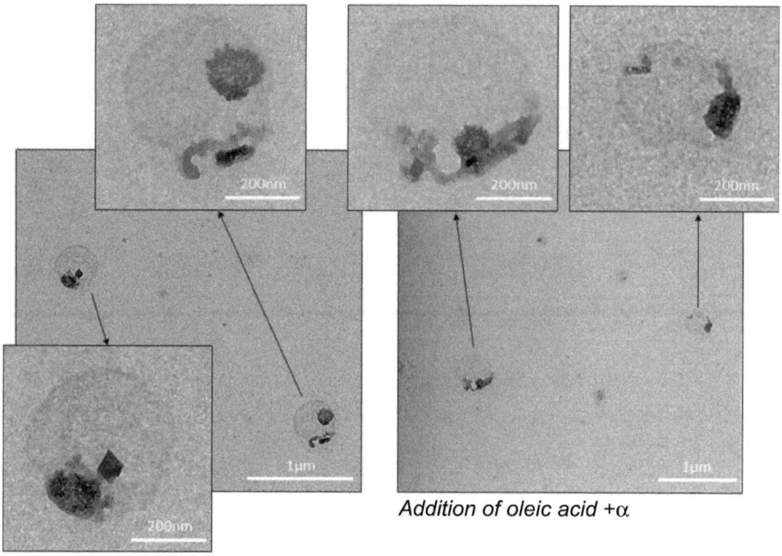

Photo 6.2 TEM image of oleic acid + α added UFB.

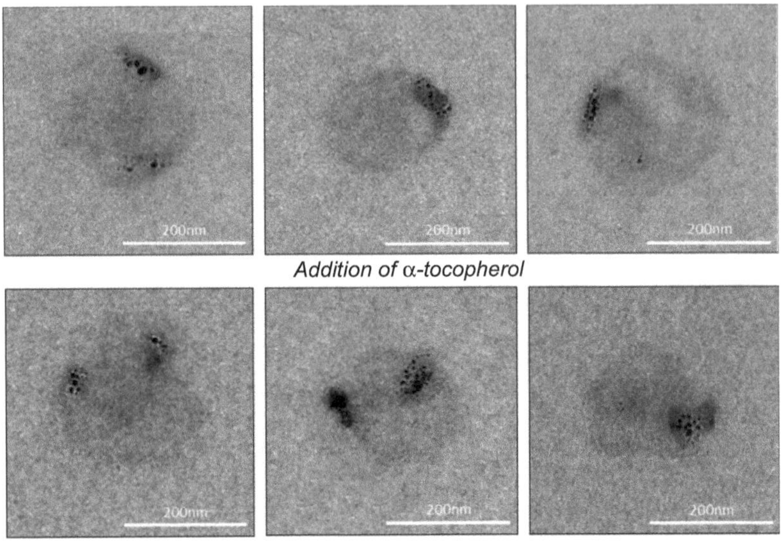

Photo 6.3 TEM image of α-tocopherol added UFB.

Photos 6.1 to 6.3 also differ in the texture of the adhering black substance because the additives are different. Also, the particle size of the UFB produced is different. It is reported that this is due to the difference in hydrophobic potential depending on the additive [1]. The bubbles in Photo 6.1 is 87–169 nm in diameter. Since the liquid phase depth is 500 nm, the bubble concentration becomes 1.8×10^{12}/mL when calculated from this image. The UFB concentration when this sample is measured by the nano tracking analysis method is 5.2×10^{10}/mL, and it is necessary to clarify the difference of the number concentration by the measurement method in the future.

In the TEM image, it looks like air bubbles, but it is measured using a resonance mass measurement method (Archimedes manufactured by Malvern) to obtain confirmation. The resonant mass measurement method is characterized by the fact that the resonance frequency varies with the particles passing through the MEMS sensor and laser-reflected light from the cantilever is detected by a photodiode detector to measure the resonance frequency, whereby bubbles and solid particles can be distinguished and measured. Specifically, since the cantilever with heavy particles vibrates slowly, they are recognized as solid particles, and the cantilever with light particles vibrates faster, so they are recognized as bubbles. Measurements were made on both micro- and nano-sensors with flow path cross section of 8 × 8 μm and 2 × 2 μm, respectively. In the case of a microsensor, since the measurement range is 0.5 μm or more, the measurement result with a higher resolution by the nano-sensor is shown in Fig. 6.17. As a result, a high positive concentration was detected within the range of 0.1 to 0.5 μm. In addition, the value of UFB number concentration measured by the resonant mass measurement method is 1/100–1/1000 of that of the nano tracking analysis method. The detected particles are almost counted as bubbles, but since hydrophobic organic materials such as oleic acid are lighter than water, there is a possibility of counting particles of oleic acid alone. However, in the TEM image, particles adhering to the bubbles could be confirmed, and a particle which can be regarded as oleic acid alone could not be confirmed. From these facts, it was confirmed that at least the particles confirmed in the TEM image are bubbles. Vitamin-E-added UFB water was also measured, but the same result was obtained. The measurement data provider is Fine Bubble Technology Office, National Institute of Technology and Evaluation (NITE).

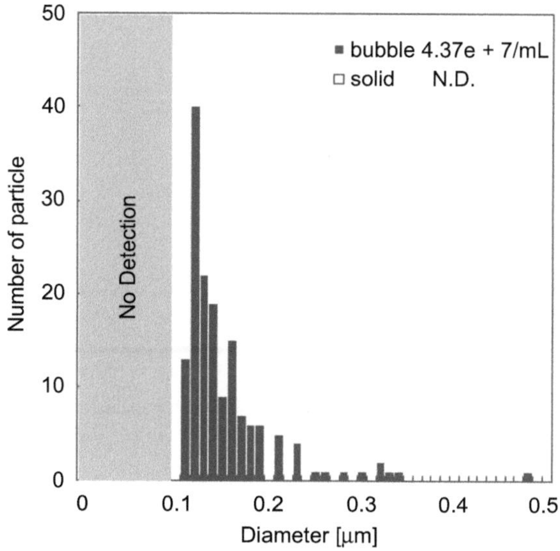

Figure 6.17 Particle size distribution of oleic acid added UFB measured by Archimedes.

Furthermore, the defoaming experiment of UFB to which oleic acid was added was performed using an ultrasonic wave. The experimental device configuration is shown in Fig. 6.18 and the experimental results are shown in Fig. 6.19.

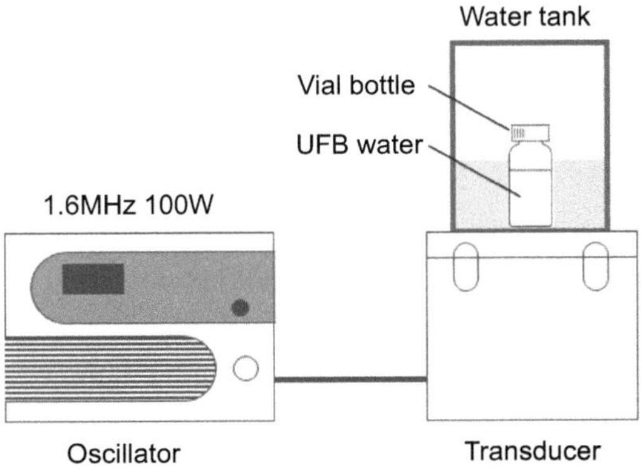

Figure 6.18 Equipment configuration of ultrasonic defoaming experiment.

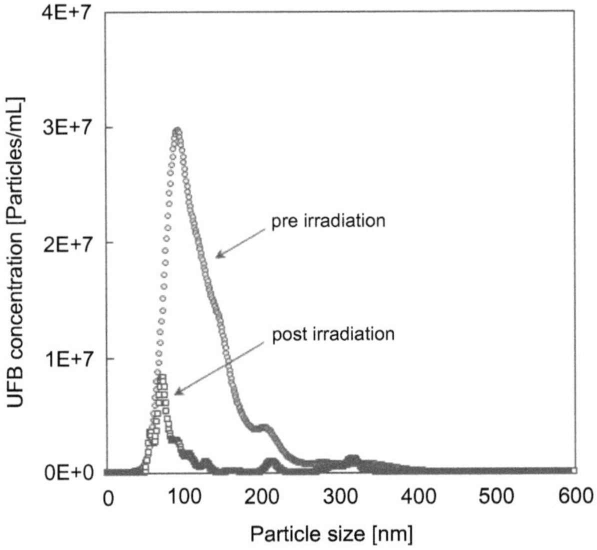

Figure 6.19 UFB defoaming by ultrasonic irradiation.

A high frequency of MHz is needed to defoam UFB with ultrasonic irradiation [13]. In this experiment, ultrasonic waves were irradiated at 1.6 MHz–100 W for 10 min. As a result, the UFB concentration before ultrasonic irradiation was 2.3×10^{10}/mL but decreased to 2.8×10^{9}/mL after irradiation. This indicates that UFB to which oleic acid is added is a bubble.

6.5 Consideration on UFB Stabilization Mechanism

As described earlier, in this experiment, the following results were obtained:

- UFB generation concentration depends on TOC concentration in liquid.
- UFB is stabilized even if it is not supersaturated.
- Suppression of extinction of UFB by addition of organic material.

From these facts, it was found that organic materials act on UFB to suppress disappearance and the UFB is stabilized even if the condition is not such that air bubbles tend to be generated due to the supersaturated state of the air and the gas is hard to dissolve. In one of the mechanisms, bubbles are stabilized by suppressing the out flux of gas from the bubbles into the water by the organic material adhering to the bubble surface. TEM analysis and resonant mass spectrometry were carried out to confirm how organic material acts on bubble structure. As a result, it was able to confirm that organic material adheres to the bubble surface. Therefore, it became clear that the adherence of organic material to the bubble surface contributes to the stabilization of bubbles.

A dynamic equilibrium model has been proposed as a bubble stabilization mechanism [1]. A schematic diagram of the dynamic equilibrium model is shown in Fig. 6.20a. It is found that it is very similar to the bubble shape actually measured by TEM observation. Ordinarily, gas escapes from bubbles into water, but in the dynamic equilibrium model, organic material adheres to a part of bubbles and acts to collect and flow gas dissolved in water into bubbles. Since it causes the action of flowing gas into the bubbles, it is the theory that air bubbles can exist stably by balancing the in fluxing amount of gas from this organic material with the amount of gas out fluxing into water from bubbles. Also, as shown in the model in Section 6.2, it is considered that the organic material also possesses the feature of the skin model in Fig. 6.20b to suppress gas out flux by covering the bubble. Therefore, it is considered that gas dissolution is further inhibited more than the dynamic equilibrium model.

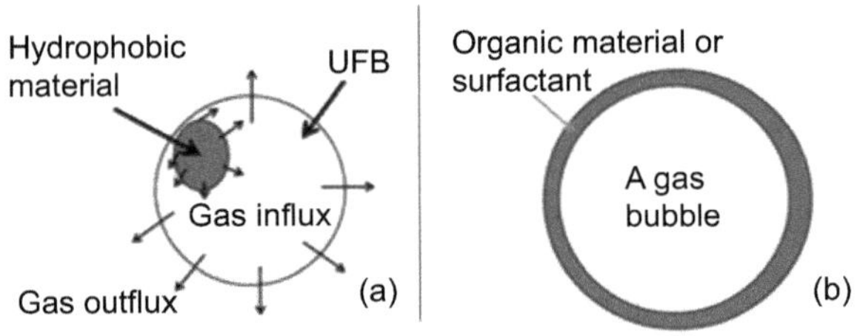

Figure 6.20 UFB image of (a) dynamic equilibrium model [1] and (b) skin model [2].

In the future, including the evaluation of the size of organic material adhering to UFB and the fraction of surface coverage by organic material, it is necessary to examine the consistency of calculation of dynamic equilibrium theory and measurement result. However, we think it significant that bubbles that support the bubble model diagram of the dynamic equilibrium theory could actually be observed. Based on the results of this experiment, we look forward to the future progress of the dynamic equilibrium theory for the foam stabilization mechanism.

6.6 Conclusion

We clarified that TOC affects the suppression of UFB extinction and found that disappearance of UFB can be suppressed by adding a small amount of organic material. Furthermore, it was confirmed by the TEM observation that organic material adhered to the surface of UFB, and organic material contributed to the suppression of UFB extinction. From these facts, the possibility of supporting the dynamic equilibrium theory as a bubble stabilization mechanism was shown.

References

1. Yasui, K. (2016). Mechanism for stability of ultrafine bubbles, *Jpn. J. Multiphase Flow,* **30**(1), pp. 19–26.
2. Yount, D. E. (1979). Skins of varying permeability: A stabilization mechanism for gas cavitation nuclei, *J. Acoust. Soc. Am.,* **65**, pp. 1429–1439.
3. Azmin, M., Mohamedi, G., Edirisinghe, M., and Stride, E. P. (2012). Dissolution of coated microbubbles: The effect of nanoparticles and surfactant concentration, *Mater. Sci. Eng. C,* **32**, pp. 2654–2658.
4. Strasberg, M. (1959). Onset of ultrasonic cavitation in tap water, *J. Acoust. Soc. Am,* **31**, pp. 163–176.
5. Takahashi, M. (2005). ζ potential of microbubbles in aqueous solutions: Electrical properties of the gas–water interface, *J. Phys. Chem. B,* **109**, pp. 21858–21864.
6. Bunkin, N. F. and Bunkin, F. V. (2013). Bubston structure of water and aqueous solutions of electrolytes, *Phys. Wave Phenom,* **21**, pp. 81–109.
7. Weijs, J. H, Seddon, J. R. T., and Lohse, D. (2012). Diffusive shielding stabilizes bulk nanobubble clusters, *Chem. Phys. Chem,* **13**, pp. 2197–2204.
8. Okamoto, R. and Onuki, A. (2015). Bubble formation in water with addition of a hydrophobic solute, *Eur. Phys. J. E,* **38**, 72.
9. Yarom, M. and Marmur, A. (2015). Stabilization of boiling nuclei by insoluble gas: Can a nanobubble cloud exit? *Langmuir,* **31**, pp. 7792–7798.
10. Yasui, K., Tuziuti, T., Kanematsu, W., and Kato, K. (2015). Advanced dynamic-equilibrium model for a nanobubble and a micropancake on a hydrophobic or hydrophilic surface, *Phys. Rev. E,* **91**, 033008.
11. Takata, T. (2010). Determination of trace impurities in highly purified reagent, *TOSOH Res. Technol. Rev.,* **54**.
12. Inazato, S., Nakazawa, E., Sakaguchi, Y., Iseki, M., Wayama, M., Nakano, K., Ominai, Y., and Konomi, M. (2016). Observation in the liquid of the ultrafine bubble by TEM. In *The Japanese Society of Microscopy, 72nd Annual Meeting,* **167**.
13. Yasuda, K., Matsushima, H., and Asakura, Y. (2019). Generation and reduction of bulk nanobubbles by ultrasonic irradiation, *Chem. Eng. Sci.,* **195**, pp. 455–461.

Chapter 7

Cleaning with Ultrafine Bubble Water

Koichi Terasaka
Department of Applied Chemistry, Keio University, 3-14-1 Hiyoshi, Kohoku-ku, Yokohama, Kanagawa 223-8522, Japan
terasaka@applc.keio.ac.jp

It is assumed that the object to be cleaned is a solid flat surface, and the dirt is oil or salt. Cleaning is completed if any dirt is removed from the surface to be cleaned. The cleaning liquid is assumed to be water-containing microbubbles or ultrafine bubbles and does not contain other detergents. Therefore, nonvolatile chemicals such as surfactants do not remain in the wastewater after cleaning, realizing a cleaning technology with less adverse effect on the environment. In this chapter, the fundamentals and a few applications are described on cleaning with fine bubble water.

7.1 Introduction

"Washing" broadly includes purification of water and wastewater, and air purification, but in this chapter, we will focus on the cleaning of solid surfaces. A wide solid surface is defined as a "surface to be

Ultrafine Bubbles
Edited by Koichi Terasaka, Kyuichi Yasui, Wataru Kanematsu, and Nobuhiro Aya

ISBN 978-981-4877-59-6 (Hardcover), 978-1-003-14195-2 (eBook)
www.jennystanford.com

cleaned," and a foreign object attached to the solid surface is defined as "dirt" [1].

Although it is difficult to clearly distinguish between "adhesion" and "adsorption," it is "adsorption" when the adhesion force between the foreign material and the solid surface is due to chemical attraction such as van der Waals force, hydrophobic interaction, or chemical bond. On the other hand, it is called "attachment" to include physical bonds by matching the surface irregularities with the shape of the foreign matter.

Solid surface cleaning is used in a wide range of applications, including semiconductor wafer finishing, painted metal surfaces, building exterior cleaning, food processing, used dish cleaning, hygiene cleaning, health and beauty, skin and hair cleaning. Conventionally, "cleaning" is often performed by combining chemical reaction or cleaning using chemical substances with "rinsing." However, a cleaning technology using fine bubbles without adding a chemical substance has attracted attention.

7.2 Interfacial Energy

A boundary between phases is generally called an "interface." However, only the interface between a liquid (or solid) and its gas (or vapor) will be called a "surface."

Figure 7.1a shows the state of molecules near gas–liquid interface at a bubble floating in liquid. Molecules present in the liquid phase can be freely repositioned, but any molecule is surrounded by other molecules. Uniform intermolecular forces are applied from all directions, so the total forces are balanced.

On the other hand, molecules existing on the outermost surface of the liquid phase in contact with the gas phase are hardly subjected to attractive force because the number of molecules in the gas phase is very few, and the resultant force of intermolecular force acts toward the liquid phase.

Therefore, molecules in the liquid phase have a smaller free energy than molecules on a gas–liquid interface. Thus, liquid minimizes the interface free energy by minimizing the number of molecules in contact with the gas–liquid interface. This is explained by the behavior that a bubble in liquid approaches sphere.

Figure 7.1b shows the state of molecules near the contact point between gas, liquid, and solid phases when a droplet is placed on a solid plate set in gas space. Molecules inside the solid phase cannot be repositioned freely because of binding with strong attractive forces such as covalent bonds and metal bonds. Molecules (atoms) on the uppermost surface of solid are in contact with molecules on the lowermost surface in a droplet, but the chemical species are not identical, so they do not balance the attractive force with the same liquid molecule.

Also molecules on the liquid interface in contact with gas make the interface close to a sphere in order to minimize the interface free energy (Fig. 7.1a).

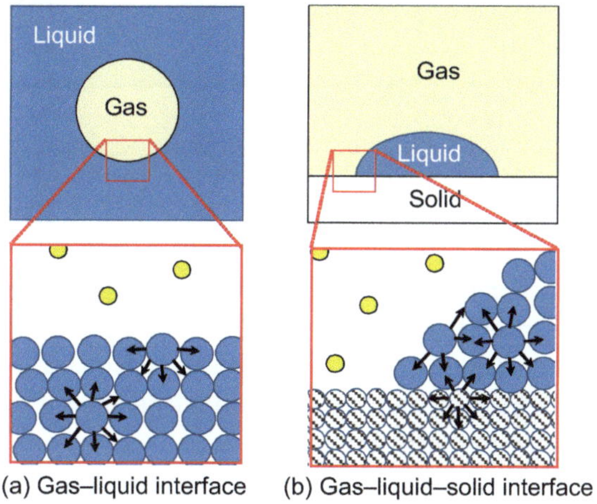

Figure 7.1 Intermolecular forces at gas–liquid interface and gas–liquid–solid interface.

7.2.1 Surface Tension and Interfacial Tension

Surface tension depends mainly on temperature and liquid species. Table 7.1 shows the surface tension σ [N/m] of several kinds of liquids. The force acting between the molecules of organic solvents, which is mainly by van der Waals force, is relatively small. On the other hand, the surface tension of gas–water is larger than that of an organic solvent because a large force due to hydrogen bonding acts between the water molecules. That is, the surface tension indicates the magnitude of the intermolecular interaction that constitutes liquid. Mercury has a large surface tension due to the strong metal bonds mediated by free electrons between atoms.

Table 7.1 Surface tension of typical gas–liquid interfaces [2]

Liquid	Formula	Gas phase	Temperature Θ [°C]	Surface tension σ [mN/m]
Hexane	C_6H_{14}	Air	20	18.4
Ethanol	CH_3CH_2OH	N_2	20	22.4
1-butanol	C_4H_9OH	N_2	20	25.4
1-hexanol	$C_6H_{13}OH$	N_2	20	26.2
Benzene	C_6H_6	N_2	20	28.9
Water	H_2O	Air	20	72.8
Metallic sodium	Na	N_2	120	199
Mercury	Hg	N_2	25	482
Iron	Fe	He	1500	1890

On the other hand, at the liquid–liquid interface as well as the solid–liquid interface, the energy of the molecules on the interface is larger than that inside the phase. For this reason, interfacial tension acts on the interface, and the interfacial free energy per unit area is represented by the interfacial tension.

7.2.2 Wettability and Contact Angle

As shown in Fig. 7.2, a small amount of liquid is placed on a smooth solid surface and a lenticular or hemispherical drop is placed on a flat surface. Evaluation by observing the contact state of the liquid phase with the solid surface is called "wettability."

When the shape of a droplet on a solid surface is viewed from the horizontal direction, the intersection between the contour line of the gas–liquid interface and the solid surface is defined as the "end point." The angle at the end point is called "contact angle," which is θ ($0° < \theta < 180°$). The surface tension of the solid (interface tension between gas/solid) is σ_S, the surface tension of the liquid (interface tension between gas/liquid) is σ_L, and the interface tension between solid and liquid (interface tension between liquid/solid) is σ_{LS}. The surface tension σ_S of the solid tries to reduce the area of the solid surface, that is, the gas/solid interface, and pulls the end point to the left.

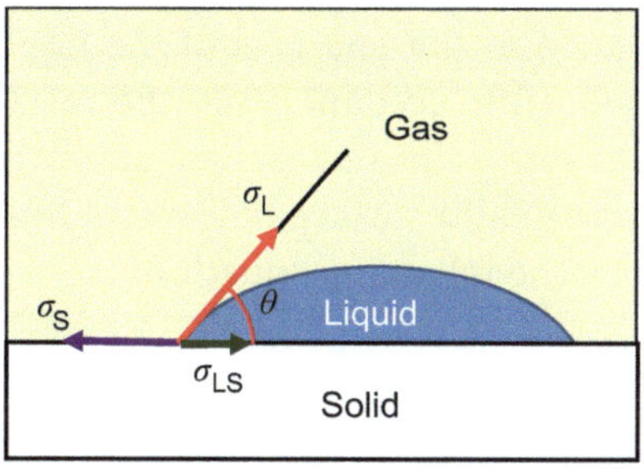

Figure 7.2 Surface tensions of gas–solid, gas–liquid, and liquid–solid interfaces.

The liquid/solid interfacial tension σ_{LS} pulls the end point to the right in an attempt to reduce the area of the liquid/solid interface.

The surface tension σ_L of the liquid acts in the tangential direction of the droplet contour in an attempt to reduce the area of the liquid surface, that is, the gas/liquid interface, but the horizontal component $\sigma_L \cos\theta$ pulls the end point to the right.

When the wettable condition is stable, the end points do not move to the right or left and remain stationary, and these three forces balance. The forces balance in the direction parallel to the solid surface, and the balance is expressed by the following Young's equation:

$$\sigma_S = \sigma_{LS} + \sigma_L \cos\theta \tag{7.1}$$

That is,

$$\cos\theta = \frac{\sigma_S - \sigma_{LS}}{\sigma_L} \tag{7.2}$$

Here cos θ is larger than 0 when $\sigma_S > \sigma_{LS}$ (when the droplets adhere, the solid surface itself decreases in energy). θ approaches 0° because of a well-wetted solid surface. On the other hand, for water droplets adhering to a water-repellent material (Teflon, etc.), $\sigma_S < \sigma_{LS}$, and θ increases and approaches 180°. A droplet becomes almost spherical.

7.2.3 Free Energy Change of Adhesion and Desorption of Dirt

When a liquid droplet adheres to a solid surface, a part of the solid surface is wetted. When considering the free energy change at a wet spot, the change in the energies of the liquid surface and the solid surface is taken into account. Since the surface tension and the interfacial tension represent energy per unit area, the free energy change ΔG, when the solid surface of the unit area causes adhesion wetting, is given by the following equation:

$$\Delta G = \sigma_{LS} - \sigma_L - \sigma_S \tag{7.3}$$

Substituting Eq. 7.3 into Eq. 7.2 gives

$$\Delta G = -\sigma_L(1 + \cos\theta) \tag{7.4}$$

When the surface tension of the liquid is large and the contact angle is small, the free energy is greatly reduced, and the solid surface is easily wetted.

On the other hand, there are frequent cases where dirt floating in the wastewater adheres to the inner surface of the container.

As shown in Fig. 7.3, a contaminant is floating in the liquid and another contaminant is in contact with the same liquid. Those contaminants contact each other and adhere to the solid surface and make a new interface.

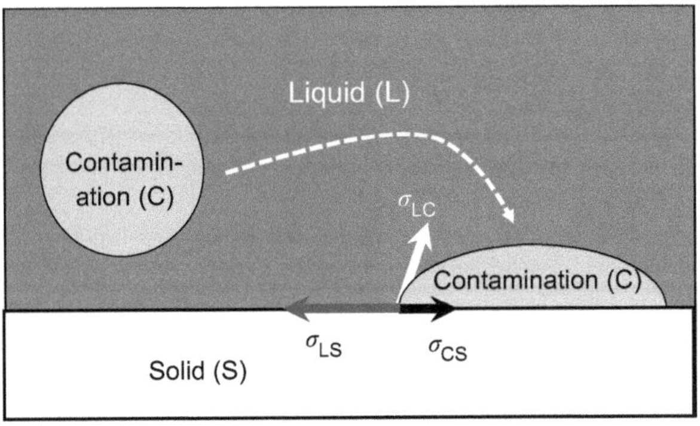

Figure 7.3 Dirt adhesion to solid surfaces in liquid phase.

Considering the change in free energy, the following equation is obtained:

$$\Delta G = \sigma_{CS} - \sigma_{LC} - \sigma_{LS} \tag{7.5}$$

where σ_{LC}, σ_{LS}, and σ_{CS} are interfacial tensions between the liquid and contaminant phases, liquid and solid phases, and contaminant and solid phases, and each value is positive.

Since the adhesion of the contaminant phase to the solid wall surface is spontaneous, the change in free energy is $\Delta G < 0$.

In order to clean the solid surface by detaching the already attached dirt phase from the solid phase interface, an energy of $\Delta G \geq 0$ is required. The energy required for cleaning is input as mechanical energy, physical energy, thermal energy, or chemical energy.

Chemical energy is brought by the addition of a detergent. When the detergent is dissolved in liquid, the dirt phase is easily dissolved or dispersed in the liquid. When the detergent is introduced, σ_{LC} is reduced, so ΔG approaches zero from negative, and the dirty phase is easily detached.

When the liquid temperature is increased by adding heat energy, the detergent surface activity increases, σ_{LC} further decreases, and the adhesion interface state becomes unstable and σ_{CS} increases. Therefore, ΔG approaches 0 from the negative side, and when it finally becomes positive ($\Delta G > 0$), the dirt phase naturally desorbs.

This is the cleaning mechanism of immersion liquid cleaning. Even if ΔG is slightly negative, the detachment of the dirt phase is promoted by mechanical energy such as the shearing force or pressure, that is, by rubbing the solid surface with a brush or by using a washing machine.

As described earlier, an appropriate combination of chemical energy, mechanical energy, and thermal energy makes cleaning more efficient.

7.3 Fine Bubble Cleaning

The cleaning of wall deposits with fine bubbles has been put into practical use. In order to quantitatively estimate or evaluate cleaning effects, the ISO has set international standards and evaluation methods [3]. In Japan, it is already widely used for washing hot water pipes, washing machines, and exterior walls of buildings.

7.3.1 Microbubble Cleaning

Cleaning oil stains on solid surfaces using microbubbles has already been put into practical use [4]. Oil adheres to a solid wall due to the negative free energy as described in Section 7.2.3, but adsorption also occurs between oil droplets having a hydrophobic surface and the surface of microbubbles [5].

As shown in Fig. 7.4, when microbubbles come close to or come in contact with oil droplets attached to the wall surface, some oil is adsorbed on the microbubble surface. Microbubbles give upward buoyancy with some oil.

A part of the oil adsorbed on the microbubbles is separated from the oil body adhering to the wall by the floating motion of the microbubbles. Some oil entrained is transported to free surface by microbubble flotation.

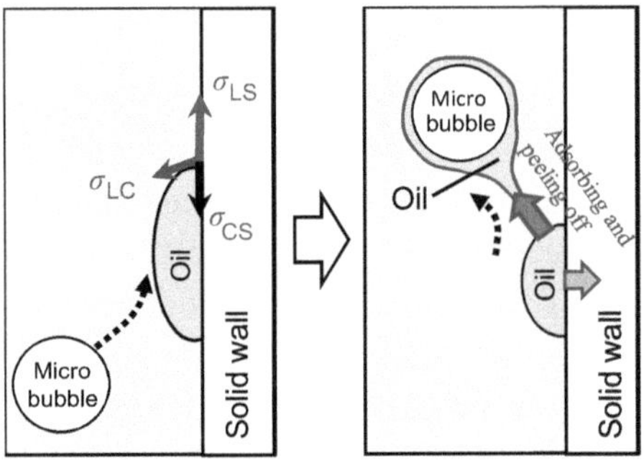

Figure 7.4 Oil contaminant adsorption to microbubbles.

Microbubbles are accompanied by a slow ascending motion in water, so if fresh microbubbles are continuously supplied into water, "microbubble water" clouded with microbubbles can be maintained. The smaller the size of the floating microbubbles, the smaller the slip velocity from the liquid velocity, so that the microbubble can approach the wall surface with liquid. The smaller the volume per microbubble and the larger the total gas volume dispersed in the container, the greater the area A_B effective for adsorption, as in Eq. 7.6, and more oil dirt can be adsorbed. In addition, the microbubbles of volume V_B coated with oil of volume V_O have an increased apparent density difference $\Delta\rho_a$, so that the buoyancy is reduced and the rising velocity in liquid is reduced or stagnated.

$$\Delta\rho_a = \frac{(\rho_L - \rho_G)V_B g + (\rho_L - \rho_O)V_O g}{(V_B + V_O)g} \tag{7.6}$$

where ρ_L, ρ_G, and ρ_O are the density of washing water, the gas density in the microbubbles, and the density of oil [kg/m^3], respectively.

After cleaning, it is necessary to remove the oil adhering to the wall surface and efficiently keep the microbubbles covered with oil away from the cleaning surface. It is often used for forced flow or continuous exchange of liquid. When reduction in the energy cost is important, natural levitation using only buoyancy (gravity) is used.

7.3.2 Ultrafine Bubble Cleaning

As a specific physical property different from microbubbles, ultrafine bubbles have static stability for a long time in water [6, 7]. Therefore, like a drug added to water, ultrafine bubbles exist at a certain concentration in water. Generally, additive components remain in liquid because liquid or solid additives are less volatile. However, when the additive is ultrafine bubbles, the encapsulated component is gaseous, so it has high volatility and low solubility in liquid and it hardly remains in the liquid after the bubbles collapse. Thus, after ultrafine bubble cleaning where ultrafine bubbles are consumed, there is almost no detergent ingredients other than dirt substances in wastewater.

Because of its characteristics, ultrafine bubble water works as a cleaning agent that does not pollute the natural environment with chemical substances. It is also a material with high affinity for food hygiene applications, hygienic environment preservation, allergy symptoms alleviation, etc.

Ultrafine bubbles are stable in water. However, when a certain amount of stimulus is applied, the stable state shifts to another stable state (dissolution of internal gas into water or change to large bubbles due to coalescence).

The most important energy that contributes to cleaning wall deposits is derived from the fluid flow. In order to desorb salt (dirty substances) adhering to a wall, the surface of the adhering material should be covered with running water. At this time, if ultrafine bubbles are contained in the running water, a higher cleaning effect can be obtained.

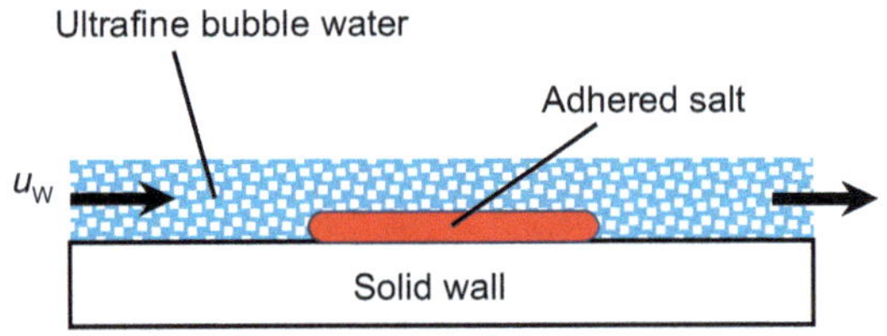

Figure 7.5 Cleaning test of salt on the wall with ultrafine bubble water.

Figure 7.5 shows a dirt model sample in which a salt of mass w_0 [mg] is fixed to a solid wall. When ultrafine bubble water is passed through the sample surface at a constant flow rate u_W [m/s], the attached salt is released by dissolution or peeling from the wall and flows downstream.

Since the mass w [mg] of the dissolved salt is obtained from the salt concentration in the washing wastewater accumulated downstream, the washing rate β [%] is

$$\beta = \frac{w}{w_0} \times 100\% \tag{7.7}$$

Unlike paste-like "soft deposits" such as biofilm and starch, when "hard deposits" such as salts are washed with running water, the progress of the cleaning rate β is as shown in Fig. 7.6.

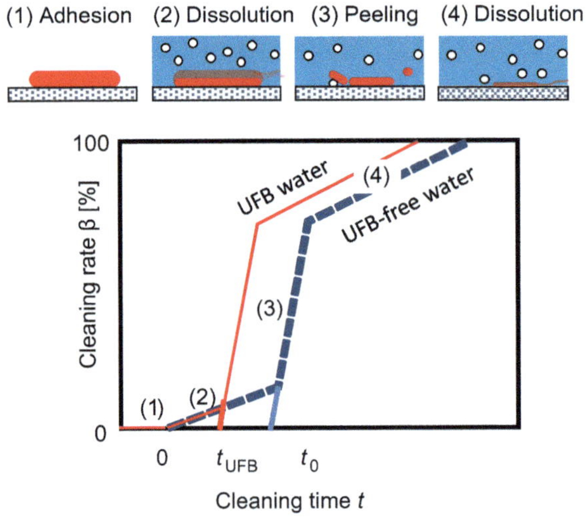

Figure 7.6 Cleaning process of salt adhering to wall surface with ultrafine bubble water.

Assuming that the time when ultrafine bubble water contacts the fixed salt is $t = 0$,

1. Dried salt is attached to the solid wall at $t < 0$.
2. Dissolution starts from the surface by contact with water at $t > 0$.
3. When $t > t_0$ or $t > t_{UFB}$, the adhering salt is separated for each small piece and most of salt is peeled off.
4. Residual salt left on the solid surface dissolves and cleaning is completed.

The cleaning process consists of the preceding four steps. At this time, if the ultrafine bubbles are contained in water, the starting time of the peeling step (3) is promoted, and as a result, the cleaning is completed in a shorter time with ultrafine bubble water.

Figure 7.7 shows a hypothesis of the cleaning mechanism of wall-attached salt with ultrafine bubble water [8].

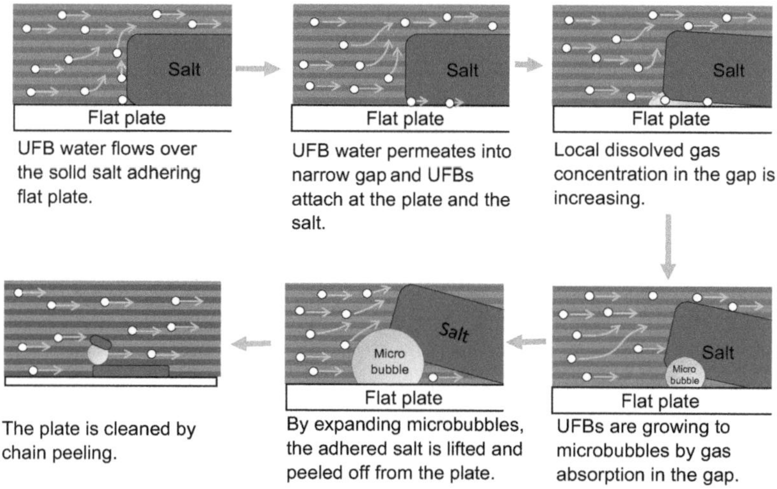

Figure 7.7 Hypothesis of removal mechanism of wall-attached salt by flowing ultrafine bubble water.

The flowing ultrafine bubble water permeates into the void between the solid wall and the attached salt. The ultrafine bubbles trapped in the gap are stable but start expanding to become microbubbles due to the supply of dissolved oxygen supersaturated in the ultrafine bubble water. Microbubbles are larger than the gap between the solid wall and the adhered salt, so some of the solid salt is jacked up. As a result, water also flows into the expanded void, destroying some of the solid salt. In this way, the solid wall with adhered salt is cleaned by ultrafine bubble water.

The cleaning mechanism using ultrafine bubbles utilizes the high permeability of ultrafine bubbles accompanying water and the metastability of ultrafine bubbles that shift to a stable state when a specific potential is exceeded.

References

1. Japan Food Engineering Association. (2012). *Food Engineering*, Asakura Shoten (in Japanese).
2. Chemical Engineering Society of Japan. (1978). *Chemical Engineering Handbook*, 4th revised edition (1978). Edited by Chemical Engineering Society of Japan, Maruzen.
3. Aya, N. (2015). Basics of Fine Bubbles and Application to Cleaning, Special Issue Application to Cleaning Fine Bubbles, Industrial Cleaning, Japan Industrial Cleaning Council, No. 16.
4. Yanase, S. and Matsuura, K. (2015). Recent development of the micro-bubble science, *Nagare*, **34**, pp. 355–362.
5. Japan Society of Mechanical Engineers. (2009). Forefront of Mechanical Engineering 3, Forefront of Micro Bubble, Kyoritsu Publishing.
6. Toray Research Center. (2015). Fine Bubbles-Fundamental, Practical and New Development of Micro/Nano Bubbles, Research Department of Toray Research Center.
7. Micro Bubble (Fine Bubble) Mechanism/Characteristic Control and Actual Application Points, Joho Kikou Co., Ltd., (2015)
8. Terasaka, K., Himuro, H., K. and Ando, Hata, T. (2016). Introduction to Fine Bubble Technology, Nikkan Kogyo Shimbun (in Japanese).

Chapter 8

Biological Effects and Applications of Ultrafine Bubbles

Takashi Kawasaki
National Institute of Advanced Industrial Science and Technology (AIST), Osaka 563-8577, Japan
takashi-kawasaki@aist.go.jp

The application of ultrafine bubbles (UFBs) is progressing; however, there are not many reports that have proved their effects based on scientific evidences. This chapter summarizes the effects of UFB on living organisms, such as plants or animals, based on scientific reports of UFB biological applications. In addition, unlike the UFB produced by a UFB generator, the plasmonic nanobubbles generated by the irradiation of laser on gold particles are discussed with respect to their generation principles and application examples.

8.1 Introduction

Ultrafine bubbles (UFB) have been practically used in various fields, such as antibacterial and sterilization applications, and for

Ultrafine Bubbles
Edited by Koichi Terasaka, Kyuichi Yasui, Wataru Kanematsu, and Nobuhiro Aya

ISBN 978-981-4877-59-6 (Hardcover), 978-1-003-14195-2 (eBook)
www.jennystanford.com

promoting the growth of plants or animals such as aquatic lives. However, there have been few reports on the scientific evidences of activities of UFB, or some reports have only simply described the results from case studies. Therefore, the mechanisms of the actions remain unclear. This chapter reports the situation of the biological activities and applications of UFBs in living organisms, referring to previous papers, most of which tried to elaborate the action mechanisms. This chapter discusses three topics: "Effect of UFB on plants," "Effects of UFB on cells and organisms of animals," and "Plasmonic nanobubble." The first and second topics include biological activities and applications of UFB-containing water (UFB water) in cells and organisms, where UFB water, produced through a UFB generator, was used. These topics describe studies suggesting the usefulness of UFBs supported by scientific reports. On the contrary, the final topic describes a different category of nanobubble, plasmonic nanobubble (PNB), which is a conditionally producible nanobubble from plasmonic nanoparticles. PNB has quite different properties from UFB generated from the generator but is promising in diagnostics and therapy.

8.2 Effects of UFBs on Plants

8.2.1 Promotion of Germination and Sprout Growth by Oxygen UFB Water

Some previous investigations reported that UFB water could accelerate the growth of various species, including plants [1–3]. In fact, UFB water, especially comprising oxygen (O_2) or air, has been applied to promote growth of crops in plant factories. However, it is not sure what kind of vegetables, which developmental stage, or what kind of administration conditions are appropriate for the application of UFB in practical uses, because the action mechanisms of UFB remain unclear in most cases.

Previous papers reported that the exogenous reactive oxygen species (ROS) are generated by micro-nanobubbles (MNBs) in the presence of dynamic stimuli such as ozone, strong acid, copper, and short-wavelength UV irradiation [4–7]. Whereas Liu et al. recently reported a new characteristic of UFB (they described as nanobubble

(NB)) that UFB water could continually produce small amounts of ROS in the water without any stimuli [8] (Fig. 8.1). ROS was thought to be harmful metabolic products that cause oxidative damage to lipids, proteins, and nucleic acids [9]. However, researches have highlighted new roles for ROS as important physiological regulators of cellular signaling pathways. This means that the appropriate concentration of ROS will have positive effects on biological functions, such as growth of plants and animals [10, 11], although the higher concentration must damage and destroy cells. Potocky et al. detected extracellular ROS at the tip of growing cultured tobacco pollen tubes [12]. Suppression of NADPH oxidase (NOX) gene expression by specific antisense oligodeoxynucleotides (ODNs) in tobacco pollen tubes lowered NOX activity and inhibited pollen tube growth. The NOX inhibitor, diphenylene iodonium chloride, and ROS scavengers also inhibited growth and ROS formation in tobacco pollen tube cultures. The growth inhibition caused by NOX antisense ODNs was rescued by exogenous hydrogen peroxide (H_2O_2) [12]. Then, it was reported that an exogenous source of ROS, as well as endogenous source, could contribute to the formation of intracellular ROS [13]. These results suggested that extracellular ROS generated from the cells of plants caused the generation of intracellular ROS and regulated the growth of plants. Therefore, the generation of ROS by UFBs offered a reasonable explanation for the physiological promotion effect of UFBs. To elucidate the mechanism of the promotion effect of ROS, it was important to make clear what kind of ROS is generated by UFBs.

Generally, ROS mainly consists of four types: superoxide anion radical ($O_2{\cdot}-$), hydrogen peroxide (H_2O_2), hydroxyl radical (·OH), and singlet oxygen ($^1O_2$), which are metabolic byproducts in both plants and animals [14]. Superoxide is the first reductant formed from molecular oxygen, a precursor to other active oxygen, and reacts with nitric oxide, which plays an important role in living organisms, to eliminate its action [15]. Hydroxyl radical is one of the most reactive radicals among active oxygens, and many biological damages caused by active oxygen are attributed to hydroxyl radical [15]. Hydrogen peroxide is not very reactive and is stable at body temperature but easily decomposes by metal ions and light to form hydroxyl radical [15]. Any of them are most likely to be generated in water, which is composed of hydrogen and oxygen. However, a clear identification of them in UFB water cannot be accomplished due to technical difficulties, although several researchers had tried.

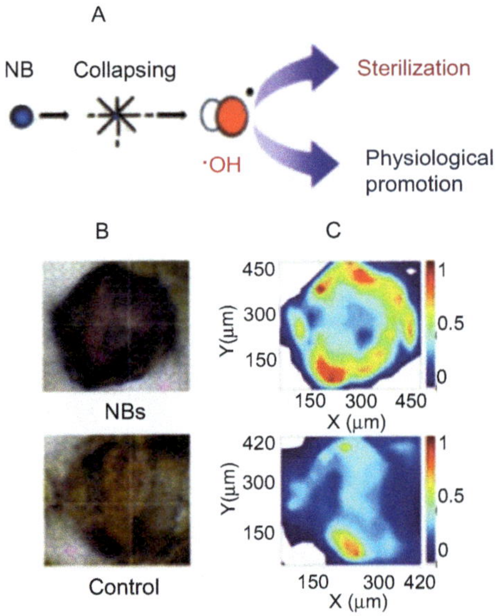

Figure 8.1 (A) Schematic of ROS generation by NBs. (B,C) Superoxide radicals distribution in barley seeds after 17 h of submerging time: the sprouting region of representative seeds germinated in two groups (distilled water, NB water) after the nitro blue tetrazolium staining process. (B) Microscopic images of the sectioned surfaces of seeds germinated in three different kinds of water. (C) Superoxide radical distribution in samples. The *x*- and *y*-axes show the locations of the measurement cycles. The measured values are the differences between absorbance values at 560 and 700. (This figure is reprinted from Ref. [8].)

Liu et al. found that seed germination of crops and their growth were promoted in UFB water made with oxygen or air [16–18] and thought that it was caused by ROS contained in UFB water. For the reason, they focused on identifying ROS in UFB water. Liu et al. developed a sensitive fluorometric method, using a sensitive fluorescent probe, 3′-p-(aminophenyl) fluorescein (APF), a Fenton reaction, a dismutation reaction of superoxide dismutase and DMSO, to identify ROS in UFB water [16] (Fig 8.2). The method enabled them to clearly distinguish four types of ROS mentioned earlier and showed that ·OH was specifically generated by O_2UFB water, whose amount positively correlated with UFB density in the water [16].

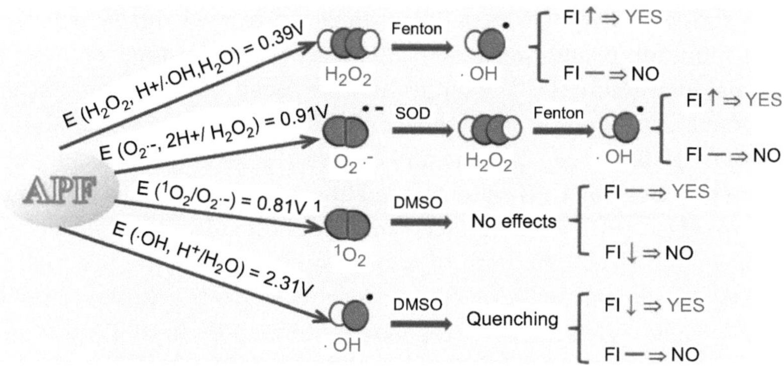

Figure 8.2 Schematic of experimental design for ROS identification (FI: Fluorescence intensity; ↑: increase; ↓: decrease; —: no change). (This figure is reprinted from Ref. [16].)

Liu et al. discussed the effects of UFB water on the germination of seeds and sprout growth based on scientific evidences [16–18]. They showed that O_2UFB water was positively effective on seed germination and the sprout growth of spinach and carrot (Fig. 8.3). They concluded that the effects were caused by hydroxy radical (·OH) produced by UFB [16]. The effects of ·OH in physiological processes depended on UFB density. When spinach and carrot seed germination tests were performed with three seed groups submerged in distilled water, high-number density UFB water (1×10^8 particles/mL), and low-number density UFB water (2×10^7 particles/mL) under similar dissolved oxygen concentrations, the germination rates and the sprout growth of spinach seeds were promoted in both UFB waters. For the carrot, the high-number density UFB water caused negative effects on hypocotyl elongation and chlorophyll formation (Fig. 8.3g), although the germination rates were promoted by UFB water [16]. The results showed that spinach and carrot had different sensitivities to ROS especially on sprout growth. In other words, it also showed that germination and sprout growth were regulated by different mechanisms. This results were consistent with the concept of a "biological window" or an "oxidative window" [19, 20] (that may be called "hormesis theory" [21]), in which a moderate level of ROS will play a positive role in physiological processes such as growth, whereas too many ROS could destroy cells and produce pathological effects by their oxidative activity. Liu et al. also analyzed a mechanism of barley germination promotion by O_2UFB water from the point of view of gene expression [17]. Comprehensive gene expression analysis about barley seeds showed that the UFB water induced expression of genes related to cell division and cell wall loosening. The exogenous ·OH produced by UFB water caused upregulation of genes related to peroxidase (POD) and reduced nicotinamide adenine dinucleotide phosphate (NADPH), both of which could catalyze the production of $O_2\cdot-$, followed by increasing endogenous $O_2\cdot-$ in the barley seeds [17]. They suggested that exogenously added ROS in UFB water and the consequently increased endogenous ROS played important roles in plant growth.

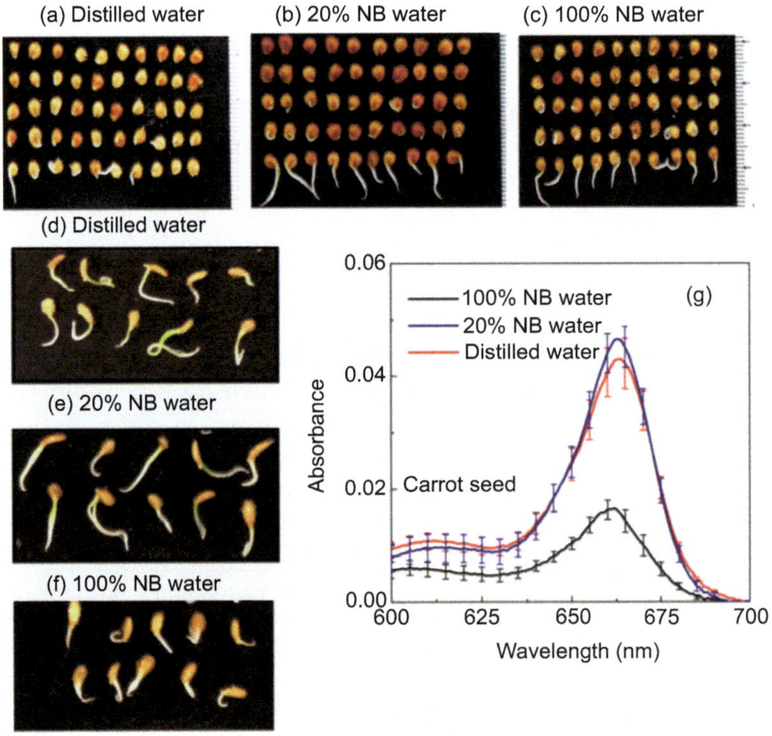

Figure 8.3 Effect of NB number density on the hypocotyl elongation and chlorophyll content of carrot seeds and spinach seeds. Fifty spinach seeds are all shown in panels (a–c). Ten carrot seeds in (d–f) are representatives of 50 seeds. (g) The absorbance of chlorophyll content of carrot seeds submerged in three kinds of waters. The error bars show the standard errors of three parallel samples. (This figure is reprinted from Ref. [16].)

The physiological importance of endogenous $O_2{\cdot}-$ in the germination of seeds was also supported by an experiment using hydrogen UFB (H_2UFB) water [18]. H_2UFB water is known to have a reducing activity, having advantages as an antioxidant that it is too mild to remove physiologically important ROS [22] and has a very small size, enough to easily penetrate biomembranes and diffuse into the cytosol [23]. However, for practical use as an antioxidant, there has been primary restrictions such as low solubility and short residence of hydrogen in water. H_2UFB water has greatly improved such restrictions by increasing dissolution and retention time of hydrogen in water [24]. Liu et al. showed antioxidant effects of H_2UFB water removing ·OH, ClO-, and ONOO- from water [18]. H_2UFB water also eliminated $O_2{\cdot}-$, a physiologically important ROS, in vivo and in vitro, contrary to the results of other researchers [22]. However, H_2UFB water failed to remove other physiologically important ROS such as H_2O_2 and NO. When barley seed germination tests were performed to study the antioxidant effect of H_2UFB water on ROS generation in vivo, the water eliminated endogenous $O_2{\cdot}-$ in seeds and inhibited germination [18]. These results supported that endogenous ROS played key roles in physiological processes.

Antioxidants have been thought to be useful in reducing the oxidative effects of ROS, but it can also adversely affect living organisms by eliminating moderate levels of ROS. It is important to maintain the oxidant–antioxidant balance in order to keep healthy conditions. For effective uses of H_2UFB water as an antioxidant, it must be carefully considered how to use, according to the intended use.

The studies on the effect of UFB water here provide a deeper understanding of the physiological promotion effects of UFB water on living organisms. Since the sensitivity to UFB water would be different among many kinds of plants, quantitative analysis and preliminary consideration are required for individual cases. This will also lead to a systematic use of UFB water depending on the kind of plants in the field of plant cultivation.

8.2.2 Promotion of Crop Growth by O_2UFB Water

Recently, UFBs have been used to promote crop growth. Previous papers reported that air-, oxygen-, and nitrogen-saturated nanobubble waters, when used for irrigation, have been demonstrated to promote the growth and improve the yield of crops, including brassica campestris, lettuce, carrot, fava bean, and tomato [2, 25].

A simple idea to explain the effect of UFB is that UFB works indirectly on crops by improving soil oxygation. Soil oxygation can improve plant root growth and nutrient uptake by supplying enough oxygen required for root respiration and energy generation. Wu et al. reported that O_2UFB water resulted in promoting the growth and increasing production of tomato by improving soil oxygation [26]. They focused on the relationship between organic fertilizer mineralization and crop growth. They showed that UFB water, made out of pure O_2/air (v/v = 1:1), increased the contents of plant-available nitrogen and phosphorus, soil enzyme activities, such as β-1,4-N-acetyl-glucosaminidase, phosphatase, α-1,4-glucosidase, β1,4-xylosidase, peroxidase, and phenol oxidase, and improved the soil microbial biomass, activity, and diversity, although some of these were similar to results with traditionally aerated water made out of pure O_2/air (v/v = 1:1). Finally, the UFB water increased the yield of tomatoes much more than the traditionally aerated water and control water. These results suggest that soil oxygation is one of the important factors in improving soil for crop cultivation, and O_2UFB water has significant effects for the purpose.

8.3 Effects of UFB on Cells and Organisms of Animals

8.3.1 Maintenance of Cells or Tissues in Animals by Delivery of Oxygen

Oxygen is indispensable for living cells or organisms; therefore, the delivery of oxygen to cells is essential to maintain the functions of cells and organisms. An experiment in vitro could directly evaluate the effects of components, including UFB, in the medium. A simple assay system using isolated cells showed that O_2UFB was effective for the increase in oxygen partial pressure (P_{O2}) in saline or blood [27]. Thus, O_2UFB may be potentially effective on the oxygenation of hypoxic tissues, for example, in infection control, for anticancer treatment, and for treatment of other ischemia.

O_2 or air UFB may have various good effects on the functions of cells in tissues or organisms through suppling oxygen. Ivannikov et al. evaluated the effects of O_2UFB water on the functions of rodent diaphragm in electrophysiological experiments [28]. O_2UFB water evoked increased neuromuscular transmission efficiency, less fatigue, and quicker recovery from fatigue than in normal medium. The effects on fatigue in the neuromuscular transmission and recovery of muscle force were greater with O_2UFB than hyperoxic solution. It is considered that the prolongation of oxygen lifetime in solution by the application of UFB technology contributed to these effects.

A paper reported the good effects of O_2UFB water even in vivo. Kanda et al. examined the effects of O_2UFB as a component of artificial cerebrospinal fluid (CSF) [29]. Spinal cord ischemic injury (SCII) after thoracoabdominal aortic aneurysm repair is a serious complication associated with cardiovascular surgery. However, effective countermeasures against SCII have not yet been established. Based on a report from Lips et al. [30], Kanda et al. formulated the hypothesis that the spinal cord tissue exposed to ischemia may consume O_2 from the CSF, and cellular respiration in the spinal cord may be preserved, at least to some extent, with O_2 transport via P_{O2} gradient. To evaluate the effect of CSF oxygenation for the prevention of SCII after infrarenal aortic occlusion in a rabbit model, they used the artificial CSF oxygenized with O_2UFB. The CSF oxygenated by O_2UFB kept P_{O2} high level, which was associated with improved neurologic function. Preservation of spinal neurons (anterior horn neurons) was also confirmed by histopathologic analysis with significant reduction in degenerated neurons, greater than that by nonoxygenated CSF. In this way, CSF oxygenation by O_2UFB can exert a protective effect against SCII.

These results suggested that oxygenation by the administration of O_2UFB was effective on improvement in cell functions not only for isolated cells but also for tissues and even for organisms. Oxygen is essential for living cells or organisms; therefore, a stable supply of oxygen is an important function of O_2UFB water, which would be expected to apply in a wide range of fields, such as agriculture, fishery, and health/medical care. It was reported that the growth of mouse and fish, including sweetfish and rainbow trout, was promoted by the administration of O_2UFB water [2]. Even these results might be due to oxygenation by UFB water, for example, by means of improvement in intestinal environment.

8.3.2 Stimulation of Cells by Ozone UFB Water

The strong bactericidal activity of ozone is due to its strong oxidizing activity. Therefore, ozone water has been expected to be applied in medical fields. However, ozone has very short half-life (30 min) in water, so it is not suitable for practical use. UFB technology combined with ozone UFB (OUFB) water has changed the situation [31]. In OUFB water, the lifetime of ozone was prolonged, and the oxidizing ability remained stable, more than six months in an electrolyte solution stored under a protected condition against UV rays. OUFB water has recently been used as an antiseptic in the treatment of periodontitis [32]. Another study investigated the utility of OUFB water for the treatment of *Helicobacter pylori* in vitro [33]. The bactericidal action was thought to be exerted by the oxidizing effect of dissolved ozone. On the other hand, there has been no report about the effects of OUFB water on normal cells in organisms. Ozone could also damage cells due to oxidative stress by means of production of ROS. It is important to investigate the action of OUFB water on normal cells, to accompany the application of OUFB water to organisms for purposes such as microorganism elimination. Leewananthawet et al. reported that OUFB water may induce oxidative stress response in human primary periodontal ligament fibroblasts (hPDLFs) [34]. They found that OUFB water had stimulated ROS production and activated the genes of hPDLFs, such as c-Fos, a major component of the transcription factor activator protein 1 (AP-1), and nuclear factor erythroid 2 (NF-E2)-related factor 2 (Nrf2), which possessed a high sensitivity to oxidative stress. This means that these factors should trigger the expression of downstream genes. RNA sequencing analysis showed that the numerous genes involved in oxidative stress responses or mitogen-activated protein kinase signaling pathway (controls a wide variety of cellular processes such as cell proliferation, growth, differentiation, transformation, and apoptosis) were upregulated after OUFB water treatment. These results suggested that cell signaling was triggered by oxidative stress produced by OUFB water in normal cells. Aqueous ozone is thought to enter cells, generate ROS, and consequently activate the relating genes. As noted earlier, ROS not only acts disturbingly, but may also serve as a regulator of cells when generated at moderate concentrations. This study may bring a new insight into the molecular mechanisms underlying the cellular responses to ozone in OUFB water and may be a base of pioneering of new application of OUFB water to cells or organisms.

8.3.3 Cancer Radiotherapy

Radiation therapy is used to eliminate cancer cells and completely cure cancer or, in combination with surgery, to prevent recurrence. However, generally, recurrent tumors in the radiation field rarely lead to cure after re-irradiation. Therefore, it is important to overcome the resistance of cancer cells on the treatment of patients. In the tumor microenvironment, consisting of malignant tumor and its surrounding stroma, hypercellular proliferation often causes a lack of oxygen due to increased distance from blood vessel, falling into hypoxia. Cancer cells adopt to hypoxic environment. Hypoxia is a characteristic feature of most tumors and known to contribute to chemoresistance, radioresistance, angiogenesis, vasculogenesis, invasiveness, metastasis, resistance to cell death, altered metabolism, and genomic instability [35–37]. Therefore, hypoxia is also considered to be an important target to treat cancers, providing some strategies, for tumor-selective therapy, including prodrugs activated by hypoxia and hypoxia-specific gene therapy.

It has been reported that cancer cells acquire radiation resistance under hypoxic condition, by means of accumulation of hypoxia inducible factor-1α (HIF-1α), one of the most important regulators in oxygen homeostasis and hypoxia adaptation [38–40]. HIF has been reported to be associated with malignancy in many types of cancers [41–43]. Genetically, HIF regulates expression of many genes, including genes that promote malignant transformation of cancer cells. Iijima et al. examined the effects of single-nanometer sized O_2UFB water on the hypoxia-induced radiation resistance and HIF-1α expression of cancer cells in vitro [44]. O_2UFB water suppressed the hypoxia-induced expression of HIF-1α in EBC–1 lung cancer and MDA–MB–231 breast cancer cells. The use of O_2UFB water also significantly reduced the hypoxia-induced radiation resistance of these cells, compared to the use of normal medium. The use of nanobubble media did not affect the viability and radiation sensitivity of these cancer cells, or the noncancerous cell line under normoxic conditions.

Similar effects of O_2UFB water have recently been shown even in vivo. Mahjour et al. evaluated the antitumor effects of O_2UFB water in cancer-bearing mice [45]. The administration of O_2UFB water to the mice significantly decreased the tumor size compared to normal water administration. The mRNA level of p53, one of the tumor suppressor genes, in the tumor cells of O_2UFB water group was found to be 36-fold higher than that in the control (normal water) group. The expression of HIF gene was significantly lower in the tumor cells of O_2UFB water group than in the control group. Interestingly, these effects could be observed just by drinking the O_2UFB water. It is unlikely that UFB taken up orally would be absorbed from the gastrointestinal tract. These results suggested that exogeneous O_2UFB may contribute to effective oxygenation of the cancer cell environment in the body.

Iijima et al. proposed a unique idea that single-nanometer sized O_2UFB could penetrate into cells in vitro, so that they used the water containing the UFB [44]. Given the characteristics due to smaller size, single-nanometer-sized O_2UFB may be able to show more effective anticancer activity in experiment in vivo.

These reports suggested that the oxygenation effects of O_2UFB water are promising for clinical application, such as more effective chemotherapy and more sensitive radiotherapy for cancer.

8.4 Plasmonic Nanobubbles

UFB water is produced by UFB generators based on principles such as gas dispersion by liquid shearing, cavitation, and gas solubility change. These techniques make it possible to create a large amount of bubble water and have been applied in various fields as described earlier. On the other hand, it is known that nanobubbles, called plasmonic nanobubbles (PNBs), are generated by irradiating a metal nanoparticle such as gold with a laser beam having an appropriate intensity on the basis of a very different principle [46–48]. This method is not suitable for the purpose of obtaining a large amount of bubble water because bubbles are generated depending on nanoparticles. However, since it can be conditionally generated only at a place where nanoparticles are present, it may be effective in a situation where an effect on cells in a living body is expected, for example. This section is focused on the basic principles and applications of PNB.

8.4.1 Basic Principles of PNB

Lapotko et al. recently introduced a new class of nano-agents, called plasmonic nanobubbles (PNBs) [46–48]. When metal nanoparticles such as gold, called plasmonic nanoparticles (NP), are irradiated with light, localized surface plasmons cause absorption at a specific wavelength. The absorbed light produces heat that is localized temporally and spatially in the vicinity of the particles due to photothermal conversion, and bubbles are produced by evaporation of the heated surrounding liquid (Fig. 8.4A). Since NPs have a very small volume, the amount of heat is small even though the temperature becomes several hundred degrees, so that the temperature returns to room temperature at a distance of several microns from the NPs and is quickly exhausted after quenching. PNB is a nanoscale explosive event induced by a laser pulse that occurs only when the energy of the laser pulse exceeds a threshold to cause the evaporation of the liquid around the NP. The generation mechanism of PNB has some unique physical properties. First, the generation of PNBs is accurately determined by the location of PNP. Since PNB does not appear in a space without NPs, it can be strictly controlled where it generates. Second, PNBs can be detected optically since the PNB surface strongly scatters light. The brightness of the scattered light rapidly increases with the size of the PNBs and becomes maximum when they have a maximum diameter. Therefore, the PNB may be detected with a scattered probe laser beam as an angle-specific positive signal or an integral negative signal [48, 49]. Third, the generation of a PNB occurs when the energy of the excitation laser pulse exceeds a certain threshold. This threshold energy was found to depend significantly upon the size and clusterization state of PNPs [46, 47, 50, 51]. It has been found that the threshold for the bubble generation decreases with increasing NP size. Besides, the plasmon properties of clustered NPs are different from those of individual

NPs, and clusters form bubbles equivalent to NPs of similar size. Fourth, the thermal impact of PNB on the surrounding is minimal. The low thermal conductivity of the vapor locks all the heat inside the bubble and thus prevents thermal impact on the environment. After the bubble collapse, the thermal impact is prevented by NPs that accumulate the additional thermal energy created during the bubble collapse [48]. These properties lead to on-demand threshold, tunability, selectivity, transience, mechanical (not thermal) action, and visualizability for PNB utilization.

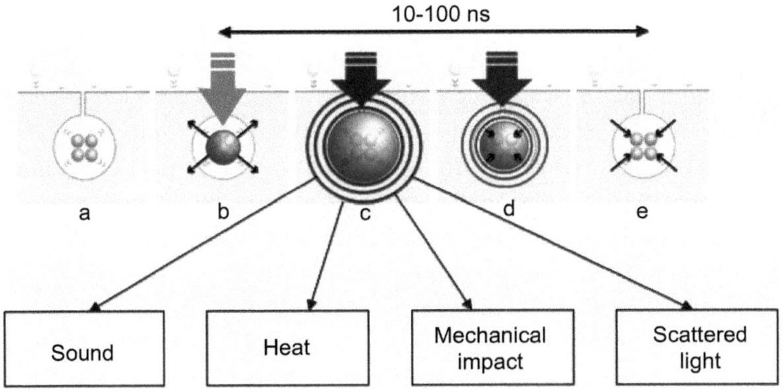

Figure 8.4 (A) The mechanism of optical generation and detection of a PNB: (a) the cluster of specifically targeted PNPs (NPs) in the target cell; (b) a short optical excitation pulse is absorbed by the PNP cluster and overheated PNPs evaporate an adjacent layer of liquid; (c) the vapor expands into a PNB that scatters optical radiation of an additional probe laser beam; (d) after reaching its maximal diameter the bubble collapses; (e) PNPs accumulate the heat that is generated during the collapse of a PNB and thus prevent the collapse-related thermal effects on the PNP environment.

8.4.2 Cell Theranostics with PNBs

In conventional medical care, diagnosis and treatment are separated from each other, which requires time to complete the treatment and impedes accuracy improvement. Recently, a new field called theranostics, which enables diagnosis and treatment of diseases in one procedure, is expected in the medical field [52–54]. The purpose of theranostics is to connect diagnosis and treatment in one procedure, shorten time from diagnosis to completion of treatment, and make them safe and efficient. The properties of PNB described above enable the development of a method for solving such medical problems. This section was focused on the applied research of PNB toward the development of methods that can contribute to theranostics.

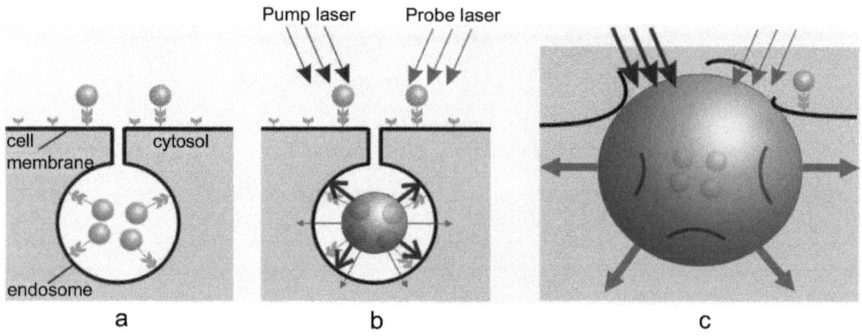

Figure 8.4 (B) Algorithm of cell theranostics with two sequential PNBs of different sizes provides the three connected steps of diagnosis, cell ablation, and immediate guidance of the cell ablation through PNB optical scattering. (These figures are reprinted from Ref. [48].)

In the current treatment of cancer, there are various problems such as recurrence due to leftover after cancer resection, unnecessary excision of normal cells, toxicity of anticancer drug/radiation to normal cells, and appearance of chemoresistant or radioresistant cancer cells; therefore, an effective treatment method has not been established yet [55]. There is a long-awaited need for the development of novel technologies that will (i) selectively detect and eliminate drug/radiation-resistant residual cancer cells to prevent recurrence, (ii) preserve the functions of normal tissues, and (iii) reduce nonspecific toxicity. An ability to unite both tumor detection and its treatment in one procedure will improve treatment efficiency [56]. Lapotko et al. have reported to establish an effective theranostic technique on the application of PNB having the characteristics described earlier [46–51, 56, 57]. They hypothesized that the effect of PNBs generated by the combination of gold NPs and optical energy would rapidly and selectively detect and eliminate drug-resistant cancer cells in a single pulse treatment (Fig. 8.4B). They further thought that the optical or acoustic detection of PNB will provide simultaneous diagnostic tumor imaging and guidance of drug delivery (Figs. 8.5 and 8.6). Utilizing models of cancers, they studied the in vitro and in vivo PNB generation, detection, and intracellular delivery of encapsulated drug (Doxil) for providing high diagnostic and therapeutic efficacy and reducing nonspecific therapeutic toxicity and the treatment time in a single theranostic procedure.

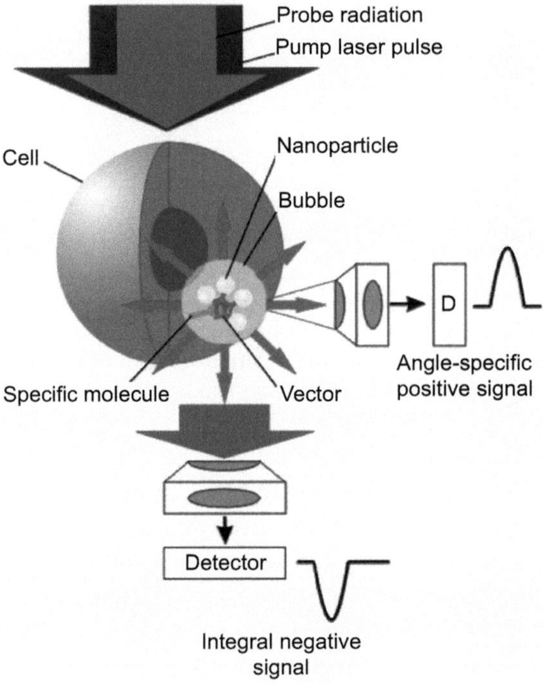

Figure 8.5 Principle of PNB amplification of optical scattering from an intracellular target: amplitude of the optical signal can be increased through the formation of a target-linked NP cluster, and through generating a transient bubble around the NP cluster; the PNB may be detected with a scattered probe laser beam (pulsed or continuous) as an angle-specific positive signal or an integral negative signal. (This figure is reprinted from Ref. [48].)

They have used gold nanoparticles (20–60 nm) conjugated with anti-epidermal growth factor receptor (EGFR) antibody, characteristic of cancer cells such as lung carcinoma cells (A549) [47, 48]. The conjugates selectively bound on the surface of A549 cells, internalized by endocytosis, localized and clustered in endosomes (Fig. 8.4). Such clusters were not observed in control (noncancer) cells having low level of EGFP expression.

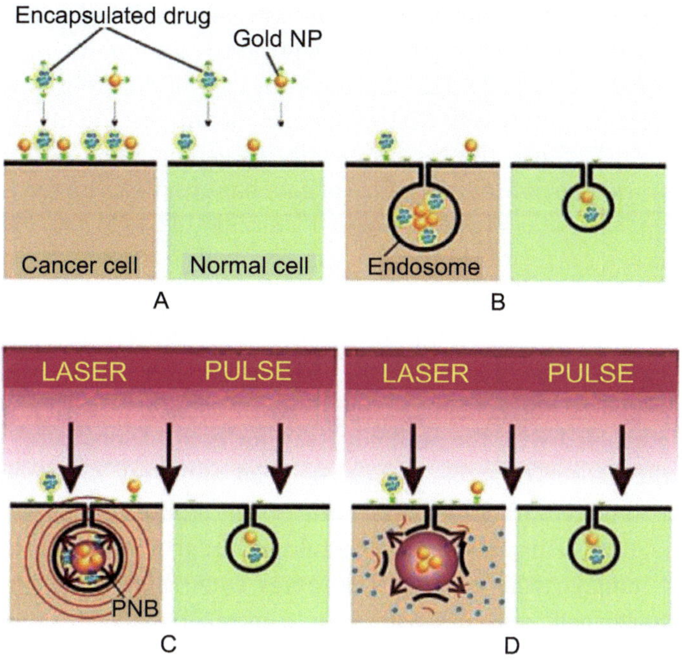

Figure 8.6 (A) Separate administration of gold NP conjugates and encapsulated drug; (B) cancer cell self-assembles mixed clusters of the drug carriers and gold NPs during receptor-mediated endocytosis; (C) diagnostic function is provided by the selective generation of PNB around the cluster of gold NPs with a single laser pulse; single gold NPs in normal (*green*) cells have a higher threshold of PNB generation and thus do not produce a PNB under a low level of laser pulse fluence; (D) selective therapeutic effect is provided by explosive localized disruption of drug carrier and endosome by PNB and the ejection of the drug (*blue dots*) into cytoplasm. (This figure is reprinted from Ref. [56].)

The radiation of excitation laser pulse to cancer cells having NP clusters caused generation of PNBs, which could be confirmed by optical (probe laser) or acoustic (ultrasound) detection [47, 48, 56]. By controlling the excitation laser intensity, it became possible to control the size of the generated PNB, and as a result, the influence of PNB on the surroundings could be controlled. As the first step of the treatment, they confirmed the specific internalization of gold nanoparticle and the generation of PNB with nanometer size and nanosecond lifetime by low intensity laser pulse irradiation (Figs. 8.4 and 8.5) (In order to detect PNBs in a tissue with low transparency, a method for detection by ultrasonic waves has also been developed [56].) In the second step, the relatively intense laser pulse irradiation transiently enlarged PNBs, consequently caused the rapid expansion and collapse of PNBs to destroy the cell membrane and kill the cells (Fig. 8.4B). Since the laser intensity threshold required for PNB generation decreases as the size of gold nanoparticles increases (the nanoparticle clusters have properties similar to single nanoparticles of the same size), the influence on the control cells can be avoided by selecting the laser intensity that generates PNBs only from clusters.

There has also been reports for specific chemotherapy combining PNB with liposomes containing an anticancer drug (Fig. 8.6) [56, 57]. By allowing the liposomes containing doxorubicin, an anticancer drug, labeled with the antibody used for gold nanoparticles to act on the cancer cells as a model of head and neck squamous cell carcinoma (HNSCC), one of the cancers with poor prognosis, at the same time using the antibody–gold nanoparticle conjugates, the gold nanoparticles and the liposomes could bind on the surface of cancer cells and be clustered into intracellular endosomes in a cancer cell-specific manner (Fig. 8.6). By the subsequent irradiation with an appropriate intensity of excitation laser pulse, the generated PNBs destroyed the liposomes as well as the endosomes, and the anticancer drug was dispersed in the cells, whereby the cancer cells could be killed (Fig. 8.6). This method has the advantages that both the concentration of the anticancer drug required for treatment and the nonspecific toxicity are low, compared with the conventional method.

The applied research of PNB for theranostics in cancer treatment, mentioned earlier, is in the experimental stage, but its usefulness is being demonstrated not only in cultured cells but also in individual animals, and it seems to be close to practical use. Unlike conventional methods using nanomaterials, which have been studied for medical applications, PNB seems to have very high safety, promptness, and efficiency in the treatment. It is considered that the theranostic method using PNB is effective in removing cancer cells in various situations. Besides, a wide range of PNB applications are envisioned, taking advantage of its characteristics, such as gene therapy.

8.5 Conclusion

UFB water made with UFB generators has been applied in various fields, including biological fields. At present, the application methods supported by scientific evidences are roughly divided into two types: one is the use of ROS, whose concentration in water is becoming measurable, and the other is controlling the supply of oxygen using UFB water made from oxygen and other gases. These effects have been reproduced at the experimental level. However, there are some problems at the practical level. First, when using a UFB generator, there is a quantitative limit to the supply. Second, it is necessary to sufficiently study the stability of UFBs when used in a solution containing various solutes. Third, for medical applications (especially when used in the body, such as in blood), the safety (sterilization) of UFB water itself must be considered. It is not easy to solve these problems, but as a direction for one solution, it may be more effective to develop and use UFB generators, having the following specifications: (i) there is no limit to the amount of production, (ii) they can prepare UFB water on various demands, and (iii) they can be easily used in a sterile state, for example, nozzle-type generators.

It is not easy to quantitatively evaluate the effect of UFBs, because UFBs are difficult to measure in various liquid conditions and their physical properties have not been sufficiently analyzed. Conversely, by establishing the measurement method and grasping the physical properties, not only can the UFB be applied more effectively, but the range of its application will be further expanded.

This chapter also focused on PNB, which can be conditionally generated in a limited space and time, as opposed to UFB water used in bulk. PNB is not suitable for use in a large amount at a time like UFB water, but their benefits may be used to establish a promising method for theranostics. In addition to such medical applications, it is expected that various usages will be developed in the future.

References

1. Park, J. and Kurata, K. Application of microbubbles to hydroponics solution promotes lettuce growth. *Hortic. Technol.* 2009, **19**, 212–215.

2. Ebina, K., Shi, K., Hirao, M., Hashimoto, J., Kawato, Y., and Kaneshito, S. Oxygen and air nanobubble water solution promote the growth of plants, fishes, and mice. *PLoS One* 2013, **8**, 1–7.
3. Ago, K., Nagasawa, K., Takita, J., Itano, R., Morii, N., Matsuda, K., and Takahashi, K. Development of an aerobic cultivation system by using a microbubble aeration technology. *J. Chem. Eng. Jpn.* 2005, **38**, 757–762.
4. Takahashi, M., Ishikawa, H., Asano, T., and Horibe, H. Effect of microbubbles on ozonized water for photoresist removal. *J. Phys. Chem. C* 2012, **116**, 12578–12583.
5. Takahashi, M., Chiba, K., and Li, P. Formation of hydroxyl radicals by collapsing ozone microbubbles under strongly acidic conditions. *J. Phys. Chem. B* 2007, **111**, 11443–11446.
6. Li, P., Takahashi, M., and Chiba, K. Enhanced free-radical generation by shrinking microbubbles using a copper catalyst. *Chemosphere* 2009, **77**, 1157–1160.
7. Tasaki, T., Wada, T., Fujimoto, K., Kai, S., and Ohe, K. Degradation of methyl orange using short-wavelength UV irradiation with oxygen microbubbles. *J. Hazard. Mater.* 2009, **162**, 1103–1110.
8. Liu, S., Oshita, S., Makino, Y., Wang, Q. H., Kawagoe, Y., and Uchida, T. Oxidative capacity of nanobubbles and its effect on seed germination. *ACS Sustainable Chem. Eng.* 2016, **4**, 1347–1353.
9. Hamanaka, R. B. and Chandel, N. S. Mitochondrial reactive oxygen species regulate cellular signaling and dictate biological outcomes. *Trends Biochem. Sci.* 2010, **35**, 505–513.
10. Brieger, K., Schiavone, S., Miller, F. J. Jr, and Krause, K. H. Reactive oxygen species: from health to disease. *Swiss Med. Wkly.* 2012, **142**, w13659.
11. Finkel, T. Signal transduction by reactive oxygen species. *J. Cell Biol.* 2011, **194**, 7–15.
12. Potocký M., Jones, M. A., Bezvoda, R., Smirnoff, N., and Zárský, V. Reactive oxygen species produced by NADPH oxidase are involved in pollen tube growth. *New Phytol.* 2007, **174**(4), 742–751.
13. Trachootham, D., Lu, W. Q., Ogasawara, M. A., Valle, N. R. D., and Huang, P. Redox regulation of cell survival. *Antioxid. Redox Signaling* 2008, **10**(8), 1343–1374.
14. Kranner, I., Roach, T., Beckett, R. P., Whitaker, C., and Minibayeva, F. V. Extracellular production of reactive oxygen species during seed germination and early seedling growth in *Pisum sativum*. *J. Plant Physiol.* 2010, **167**, 805–811.

15. Gulcin, İ. Antioxidants and antioxidant methods: an updated overview. *Arch Toxicol.* 2020, **94**(3), 651–715.

16. Liu, S., Oshita, S., Kawabata, S., Makino, Y., and Yoshimoto, T. Identification of ROS produced by nanobubbles and their positive and negative effects on vegetable seed germination. *Langmuir* 2016, **32**(43), 11295–11302.

17. Liu, S., Oshita, S., Kawabata, S., and Thuyet, D. Q. Nanobubble water's promotion effect of barley (*Hordeum vulgare* L.) sprouts supported by RNA-Seq analysis. *Langmuir* 2017, **33**(43), 12478–12486.

18. Liu, S., Oshita, S., Thuyet, D. Q., Saito, M., and Yoshimoto, T. Antioxidant activity of hydrogen nanobubbles in water with different reactive oxygen species both in vivo and in vitro. *Langmuir* 2018, **34**(39), 11878–11885.

19. Brieger, K., Schiavone, S., Miller, F. J. Jr, and Krause, K. H. Reactive oxygen species: From health to disease. *Swiss Med. Wkly.* 2012, **142**, w13659.

20. Bailly, C., El-Maarouf-Bouteau, H., and Corbineau, F. From intracellular signaling networks to cell death: The dual role of reactive oxygen species in seed physiology. *C. R. Biol.* 2008, **331**, 806–814.

21. Radak, Z., Chung, H. Y., and Koltani, E. Exercise, oxidative stress and hormesis. *Ageing Res. Rev.* 2008, **7**(1), 34–42.

22. Ohta, S. Molecular hydrogen is a novel antioxidant to efficiently reduce oxidative stress with potential for the improvement of mitochondrial diseases. *Biochim. Biophys. Acta, Gen. Subj.* 2012, **1820**, 586–594.

23. Ohsawa, I., Ishikawa, M., Takahashi, K., Watanabe, M., Nishimaki, K., Yamagata, K., Katsura, K., Katayama, Y., Asoh, S., and Ohta, S. Hydrogen acts as a therapeutic antioxidant by selectively reducing cytotoxic oxygen radicals. *Nat. Med.* 2007, **13**, 688–694.

24. Ushikubo, F. Y., Furukawa, T., Nakagawa, R., Enari, M., Makino, Y., Kawagoe, Y., Shiina, T., and Oshita, S. Evidence of the existence and the stability of nano-bubbles in water. *Colloids Surf., A* 2010, **361**, 31–37.

25. Ahmed, A. K. A., Shi, X., Hua, L., Manzueta, L., Qing, W., Marhaba, T., and Zhang, W. Influences of air, oxygen, nitrogen, and carbon dioxide nanobubbles on seed germination and plant growth. *J. Agric Food Chem.* 2018, **66**(20), 5117–5124.

26. Wu, Y., Lyu, T., Yue, B., Tonoli, E., Verderio, E. A. M., Ma, Y., and Pan, G. Enhancement of tomato plant growth and productivity in organic farming by agri-nanotechnology using nanobubble oxygation. *J. Agric. Food Chem.* 2019, **67**(39), 10823–10831.

27. Matsuki, N., Ichiba, S., Ishikawa, T., Nagano, O., Takeda, M., Ujike, Y., and Yamaguchi, T. Blood oxygenation using microbubble suspensions. *Eur. Biophys. J.* 2012, **41**(6), 571–578.

28. Ivannikov, M. V., Sugimori, M., and Llinás, R. R. Neuromuscular transmission and muscle fatigue changes by nanostructured oxygen. *Muscle Nerve.* 2017, **55**(4), 555–563.

29. Kanda, K., Adachi, O., Kawatsu, S., Sakatsume, K., Kumagai, K., Kawamoto, S., and Saiki, Y. Oxygenation of the cerebrospinal fluid with artificial cerebrospinal fluid can ameliorate a spinal cord ischemic injury in a rabbit model. *J. Thorac. Cardiovasc. Surg.* 2016, **152**(5), 1401–1409.

30. Lips, J., de Haan, P., Bouma, G. J., Holman, R., van Dongen, E., and Kalkman, C. J. Continuous monitoring of cerebrospinal fluid oxygen tension in relation to motor evoked potentials during spinal cord ischemia in pigs. *Anesthesiology* 2005, 102, 340–345.

31. Chiba, K. and Takahashi, M. Nano-bubble production method. *Jpn Pat.* 2008, P4144669.

32. Hayakumo, S., Arakawa, S., Mano, Y., and Izumi, Y. Clinical and microbiological effects of ozone nano-bubble water irrigation as an adjunct to mechanical subgingival debridement in periodontitis patients in a randomized controlled trial. *Clin. Oral Invest.* 2013, **17**, 379–388.

33. Kawara, F., Inoue, J., Takenaka, M., Hoshi, N., Masuda, A., Nishiumi, S., Kutsumi, H., Azuma, T., and Ohdaira, T. The influences of pepsin concentrations and pH levels on the disinfective activity of ozone nanobubble water against *Helicobacter pylori*. *Digestion* 2014, **90**(1), 10–17.

34. Leewananthawet, A., Arakawa, S., Okano, T., Daitoku Kinoshita, R., Ashida, H., Izumi, Y., and Suzuki, T. Ozone ultrafine bubble water induces the cellular signaling involved in oxidative stress responses in human periodontal ligament fibroblasts. *Sci. Technol. Adv. Mater.* 2019, **20**(1), 589–598.

35. Wilson, W. R. and Hay, M. P. Targeting hypoxia in cancer therapy. *Nat. Rev. Cancer* 2011, **11**, 393–410.

36. Harrison, L. B., Chadha, M., Hill, R. J., Hu, K., and Shasha, D. Impact of tumor hypoxia and anemia on radiation therapy outcomes. *Oncologist* 2002, **7**, 492–508.

37. Brown, J. M. and Wilson, W. R. Exploiting tumour hypoxia in cancer treatment. *Nat. Rev. Cancer* 2004, **4**, 437–447.

38. Ke, Q. and Costa, M. Hypoxia-inducible factor-1 (HIF-1). *Mol. Pharmacol.* 2006, **70**, 1469–1480.

39. Semenza, G. L. Hypoxia-inducible factors: Mediators of cancer progression and targets for cancer therapy. *Trends Pharmacol. Sci.* 2012, **33**, 207–214.

40. Keith, B., Johnson, R. S., and Simon, M. C. HIF1α and HIF2α: Sibling rivalry in hypoxic tumour growth and progression. *Nat. Rev. Cancer* 2011, **12**, 9–22.

41. Lu, X. and Yibin, K. Hypoxia and hypoxia-inducible factors: Master regulators of metastasis, *Clin. Cancer Res.* 2010, **16**, 5928.

42. Semenza, G. L. HIF-1: Upstream and downstream of cancer metabolism. *Curr. Opin. Genet. Dev.* 2010, **20**, 51.

43. Semenza, G. L. Molecular mechanisms mediating metastasis of hypoxic breast cancer cells. *Trends Mol. Med.* 2012, **18**, 534.

44. Iijima, M., Gombodorj, N., Tachibana, Y., Tachibana, K., Yokobori, T., Honma, K., Nakano, T., Int Asao, T., Kuwahara, R., Aoyama, K., Yasuda, H., Kelly, M., Kuwano, H., and Yamanouchi, D. Development of single nanometer-sized ultrafine oxygen bubbles to overcome the hypoxia-induced resistance to radiation therapy via the suppression of hypoxia-inducible factor-1α. *J. Oncol.* 2018, **52**(3), 679–686.

45. Mahjour, A., Khazaei, M., Nourmohammadi, E., Khoshdel-Sarkarizi, H., Ebrahimzadeh-Bideskan, A., Rahimi, H. R., and Safipour Afshar, A. Evaluation of antitumor effect of oxygen nanobubble water on breast cancer-bearing BALB/c mice. *J. Cell Biochem.* 2019, **120**(9), 15546–15552.

46. Lukianova-Hleb, E., Hu, Y., Latterini, L., Tarpani, L., Lee, S., Drezek, R. A., Hafner, J. H., and Lapotko, D. O. Plasmonic nanobubbles as transient vapor nanobubbles generated around plasmonic nanoparticles. *ACS Nano.* 2010, **4**(4), 2109–2123.

47. Lukianova-Hleb, E. Y., Hanna, E. Y., Hafner, J. H., and Lapotko, D. O. Tunable plasmonic nanobubbles for cell theranostics. *Nanotechnology* 2010, **21**(8), 85102. Erratum in: *Nanotechnology* 2016, **27**(10), 109501.

48. Lapotko, D. Plasmonic nanobubbles as tunable cellular probes for cancer theranostics. *Cancers* 2011, **3**, 802–840.

49. Wagner, D. S., Delk, N. A., Lukianova-Hleb, E. Y., Hafner, J. H., Farach-Carson, M. C., and Lapotko, D. The in vivo performance of plasmonic nanobubbles as cell theranostic agents in zebrafish hosting prostate cancer xenografts. *Biomaterials* 2010, **31**, 7567–7574.

50. Lapotko, D. Optical excitation and detection of vapor bubbles around plasmonic nanoparticles. *Optics Express* 2009, **17**, 2538–2556. *Cancers* 2011, **3**, 833.

51. Hleb, E. Y., Hafner, J. H., Myers, J. N., Hanna, E. Y., Rostro, B. C., Zhdanok, S. A., and Lapotko, D. O. LANTCET: Elimination of solid tumor cells with photothermal bubbles generated around clusters of gold nanoparticles. *Nanomedicine* 2008, **3**, 647–667.

52. McCarthy, J. R. The future of theranostic nanoagents. *Nanomedicine* 2009, **4**, 693–695.

53. Hartman, K. B., Wilson, L. J., and Rosenblum, M. G. Detecting and treating cancer with nanotechnology. *Mol. Diagn. Ther.* 2008, **12**, 1–14.

54. Picard, F. J. and Bergeron, M. G. Rapid molecular theranostics in infectious diseases. *Drug Discov. Today* 2002, **7**, 1092–1101.

55. Jemal, A., Siegel, R., Xu, J., and Ward, E. Cancer statistics, 2010. *CA Cancer J. Clin.* 2010, **60**(5), 277–300. Erratum in *CA Cancer J. Clin.* 2011, **61**(2), 133–134.

56. Lukianova-Hleb, E. Y., Ren, X., Townley, D., Wu, X., Kupferman, M. E., and Lapotko, D. O. Plasmonic nanobubbles rapidly detect and destroy drug-resistant tumors. *Theranostics* 2012, **2**(10), 976–987.

57. Lukianova-Hleb, E. Y., Belyanin, A., Kashinath, S., Wu, X., and Lapotko, D. O. Plasmonic nanobubble-enhanced endosomal escape processes for selective and guided intracellular delivery of chemotherapy to drug-resistant cancer cells. *Biomaterials* 2012, **33**(6), 1821–1826.

Chapter 9

Recent Trends in Application of Encapsulated Ultrafine Bubbles in Medicine

Katsuro Tachibana and Hiroshi Kida
Department Anatomy, Fukuoka University School of Medicine,
7-1-45 Nanakuma, Jonan-ku, Fukuoka-shi, Fukuoka-ken 814-0180, Japan
k-tachi@fukuoka-u.ac.jp

Encapsulated ultrafine bubble is an emerging field in medicine where just until recently, no one has ever thoroughly evaluated or even considered its use as a diagnostic or therapeutic tool. As microbubbles are already widely available in the clinical situation for ultrasound imaging, in that aspect, exploring the possibility of a new medical application with ultrafine bubbles could be considered the "final frontier."

9.1 Introduction

For the past three decades, encapsulated microbubbles (MBs) have been used in clinical settings as an ultrasound contrast agent. These

Ultrafine Bubbles
Edited by Koichi Terasaka, Kyuichi Yasui, Wataru Kanematsu, and Nobuhiro Aya

ISBN 978-981-4877-59-6 (Hardcover), 978-1-003-14195-2 (eBook)
www.jennystanford.com

encapsulated MBs have a diameter ranging from 1 μm to 10 μm and thin shells composed of phospholipids, albumin, protein, polymers, or galactose. MB-mediated ultrasound therapy has also attracted much attention in the research field as a future therapeutic modality. Various types of MBs and acoustic parameters have been intensively investigated in therapeutic application for many diseases. MBs can also encapsulate various drugs that could carry them to specific tissues and organs. Researchers have added modifications to the bubble shells with marker molecules that bind to targeted tissue sites, for example by coupling specific ligands. Only in these few years, the trend has moved to applying smaller-sized bubbles instead of the aforementioned MBs, creating great excitement for use in a broader range to treat difficult-to-cure diseases. This chapter focuses on recent progress of nano-sized bubble (ultrafine bubble; UFB) application in medicine.

9.2 Ultrasound Contrast Agents

Ultrasound contrast agents (MBs) for imaging have been in use mainly in the diagnosis of cardiovascular, abdominal organ diseases. The acoustic mismatch between gas and surrounding tissues when bubbles are exposed to an acoustic field creates reflection, backscatter, and harmonic responses. This results in "contrast" within the echographic image in the body, differentiating between the MB and adjacent soft tissue structures. Intravenous injection of ultrasound contrast agents can be used to identify tissue vascularization and vessel occlusion. Thus, MBs can also differentiate between benign and malignant tumors in abdominal organs such as spleen, liver, and ovaries. However, the size of each MB usually inhibits it from extravasating to surrounding tissues, thus diagnostic application is limited to the intravascular system to assess functional parameters such as red blood cell perfusion, vascularity, and flow velocity. There are currently several contrast agents available commercially, differing in gas types as air, sulfur hexafluoride, octafluoropropane, perflexane, and perfluorobutane. Moreover, the composition of the shell material can be varied to encapsulate these insoluble gases by a protein, phospholipid, or polymer. These contrast agents all have different biological and physical imaging characteristics, which may affect clinical assessment [1].

9.3 Microbubble for Drug Delivery

Microbubbles oscillate in an acoustic driving pressure field, thus inducing complex streaming in the surrounding fluid. At higher acoustic amplitudes, bubble oscillation eventually leads to asymmetrical collapse to form violent jet streaming. As a result, if a drug exists within the surrounding fluid, it will promote more drug delivery into a near-by cell or tissue. Similarly, if a drug is encapsulated inside an MB, the collapse will release the incorporated drug payload at a specific site in the body (Fig. 9.1). Tachibana et al. [2] first observed more efficient sonothrombolysis in the presence of thrombolytic agents and MB contrast agents. It was speculated that the pre-existing ultrasound contrast agent played a role as cavitation nuclei to induce synergistic effect on enhanced clot lysis and degradation of fibrin within the thrombus. This phenomenon can be classified as one of many pharmaceutical approaches often termed drug-delivery system (DDS). DDS is defined as a method in which pharmaceutical compounds are delivered to a desired tissue, cell, or organ for drug release and absorption in the body through drug carriers and physical energy. There is an urgent necessity to overcome fundamental problems in drug delivery, such as low bioavailability, poor biodistribution, lack of selectivity, and over-dosage of the drug. Researchers are continuously developing innovative DDS to solve these issues. The emerging strategy of combining ultrasound and MB is an ideal form of DDS, and the phenomenon is often termed specifically as "sonoporation" due to the fact that collapsing MBs actually induce temporary cell membrane pores made by violent jet streaming. This thus results in the injection of pharmaceutical compounds into a specific cell or through a membrane [3]. On the contrary, ultrasound contrast agents are mainly used for the diagnosis of cardiovascular, abdominal organ diseases.

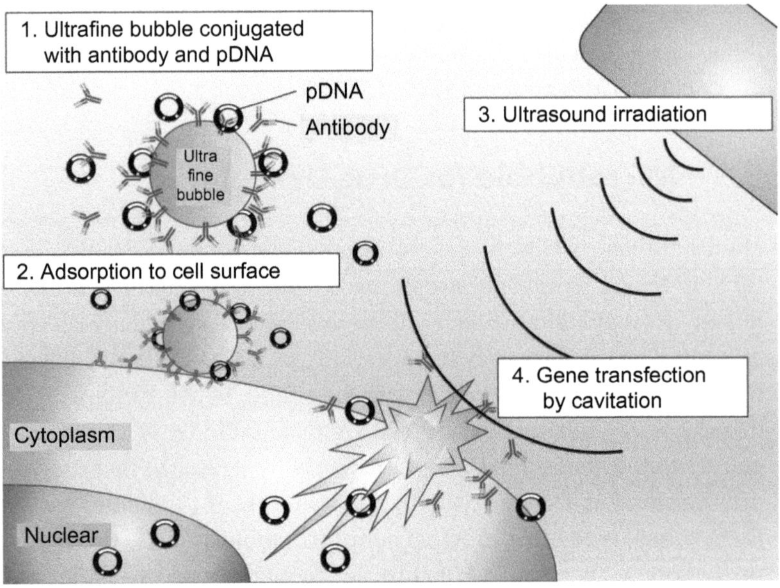

Figure 9.1 An example of a targeted UFB for sonoporation drug delivery to cancer cells in combination with ultrasound irradiation.

9.4 The Smaller the Better

There are several reasons why the trend in bubble research for medicine is rapidly shifting from micro- to nano-sized bubbles. By far the greatest reason is the fact that ultrafine bubbles (UFBs) may be able to extravasate beyond the blood vessel into the surrounding tissue. As UFBs are smaller than the conventional clinically used MB by one to two digits, the advantages are enormous from an acoustic imaging and therapeutic standpoint. The UFBs may pass freely through such tough biological barriers as the stratum corneum of the skin, blood–brain barrier (BBB), tumor vasculature, and cell membrane. In addition, UFBs can carry various compounds, for example, anticancer drugs or genes, through physiological or pathological "barriers," thus making it possible to deliver drugs to hard-to-reach sites in the body. Ultrasound is an effective adjuvant modality, which has been well demonstrated in controlling the drug release location and time. Irradiation of ultrasound at a specific timeframe is an important factor in obtaining sufficient treatment effectively for some diseases. It could be concluded that similar mechanism is also applied for UFB in a broader range in various problematic medical situations.

Several literatures have shown that UFBs can produce higher resolution ultrasound images and visualize detailed biodistribution in normal organs and malignant tissues [4]. As a novel ultrasound contrast agent, UFBs may give more favorable information to accurately diagnose malignant tumors, which would have been overlooked with conventional contrast agent echography. Although UFB ultrasound imaging technology is still in its infancy and not yet applied in humans, initial reports have demonstrated promising results in animals. Nevertheless, a large number of papers have been published just in the recent few years, showing the significance of UFB imaging compared to conventional sized MBs [5–8].

Many researchers have also reported application of UFB-mediated sonoporation for drug delivery. So far, UFB-related experimental results have shown comparable or better sonoporation drug-delivery efficiency than MBs. It can be especially noted that UFBs induce less damage to the cell membrane compared to MBs, thus leading to higher gene transfection rate in bubble-mediated ultrasound gene therapy [9]. Although there is still much discussion on the optimal method of UFB fabrication, once produced, UFBs are more robust and researchers do not have to worry of floatation of bubbles and uneven drug distribution within the solution before experimentation, which frequently resulted in varied data in the case of MBs. The stability of UFBs greatly contributes to obtaining reproducible experimental results. In addition, numerous reports have demonstrated fabrication of stable and well-characterized UFBs with various bubble shell material [10–12]. The shell usually consists of a functional targeting material permitting the drug-loaded UFBs to adsorb to a specific type of cell such as cancer (Fig. 9.1). This is an important factor especially when most anticancer agents are highly toxic; thus, UFBs can be effective only for cancer cells and, at the same time, preserve the surrounding normal tissue. Furthermore, more efficient gene transfer can take place at the targeted cell. The vast amount of technical information provided by previous reports dealing with MBs can become a steppingstone for these recently emerging UFBs for imaging and therapy.

9.5 Drug-Delivery Systems

9.5.1 Chemotherapy

Most chemotherapeutic drugs are nonselective in terms of targeting the relevant tissues, which results in adverse side effects on normal tissue. In the recent years, tumor drug targeting has evolved as a promising strategy to increase local drug concentration at a specific site in the body. As mentioned previously, a recent approach for targeting solid tumors is the application of UFBs that are preloaded with chemotherapeutic drugs and also bubble shell designed as a functional target adsorber for cancer. Furthermore, targeted drug delivery under image guidance is gaining more interest in the overall ultrasound imaging area. The use of UFBs as contrast agents with diagnostic ultrasound provides new opportunities in noninvasive image-guided drug delivery. This "diagnostic" and "therapy" linkage in one pharmaceutical component has been termed theranostics. One of the early reports on UFB theranostics was from Xu et al. [13], who synthesized and bound dual-mode poly lactic-co-glycolic acid (PLGA) UFBs for cancer targeting and imaging. A modified double emulsion method was used to encapsulate Texas Red fluorescence dye in the PLGA ultrafine bubbles. The UFB size distribution, the encapsulation efficiency, and the molecular binding efficiency were well characterized in this study.

Researchers have established methods to preload UFBs with various chemotherapeutic drugs. Figure 9.2 illustrates the procedure of a typical animal experiment where a drug of interest is injected intravenously to an animal and ultrasound is irradiated noninvasively from the skin to the target site. Most of the UFBs can be visualized by ultrasound imaging probe before collapsing the bubble with higher intensity acoustic pressure to release the preloaded chemotherapeutic drugs. Such drugs as methotrexate [14], doxorubicin [15], peptide–doxorubicin conjugate [16], folate-conjugated N-palmitoyl chitosan shell [17], apatinib-loaded lipid bubbles [18], pro-apoptotic liposomes conjugate synergistic with paclitaxel [7], porphyrin-loaded pluronic UFBs [6], and peptide–camptothecin conjugates [19] have all shown to be effective to cancer-cell culture studies and also in animal model experiments.

Although the combination of targeted shell material and drug-loaded UFBs is promising as a theranostic agent, much more optimization and safety evaluation are needed for future applications in the clinical field.

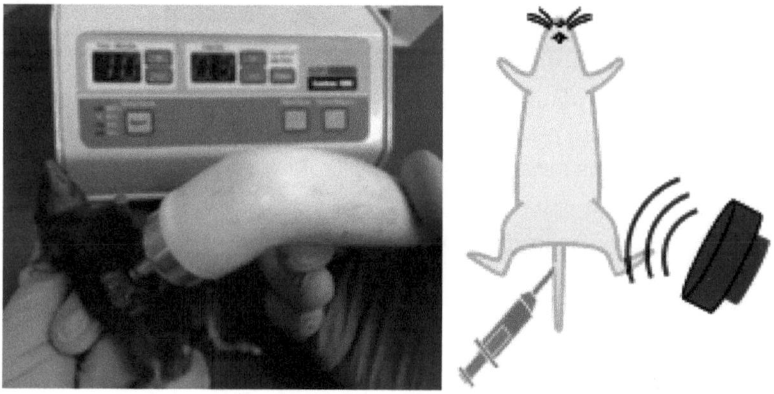

Figure 9.2 Ultrasound irradiation to a mouse after intravenous injection of drug-loaded UFBs from the tail.

9.5.2 Cardiovascular Applications

The majority of chemotherapeutic approaches of using UFBs is for malignant tumors; however, there have been several reports recently in the cardiovascular field. Cases such as monitoring patients with acute heart transplantation rejection require repeated invasive endomyocardial biopsies, and noninvasive diagnostic techniques are desperately needed. It is known that T lymphocyte infiltration is the central process of acute rejection. Ultrasound molecular imaging with T lymphocyte-targeted UFBs could be used to detect acute rejection in heart transplantation. In this study, Liu et al. [20] fabricated a UFB-bearing anti-CD3 antibody (NBCD3) or isotype antibody (NBcon), which showed significant adhesion to T lymphocytes in vitro. The acoustic signal intensity of the adherent NBCD3 was significantly higher than that of the NBcon in allograft rats, but not significantly high in isograft rats. It was suggested that ultrasound molecular imaging could be used as a noninvasive method in acute rejection detection after cardiac transplantation. Another study [21] also reported assessment of hepatic ischemia-reperfusion injury specifically, noninvasively, and quantitatively at the molecular level. A nanoscale bubble contrast agent targeting anti-intracellular adhesion molecule-1 (anti-ICAM-1) was administrated to liver ischemia-reperfusion injury rabbit model. Contrast-enhanced ultrasound images showed that the peak intensity and the time of duration of the targeted UFBs were significantly higher than nontargeted UFBs. Other cardiovascular studies have targeted atherosclerotic plaques [22], UFB vasodilator within porcine carotid arteries [23], and platelet membrane-based UFBs for targeting acute ischemic lesions for acute ischemic stroke lesions [24].

9.5.3 Bacteriological Applications

Most bacterial diseases can be treated with antibiotics; however, antibiotic-resistant strains have become a major issue especially for immunodeficient patients and the elderly. Previous work has demonstrated that ultrasound irradiation enhances the bactericidal activity of antibiotics in *Pseudomonas aeruginosa* or *Escherichia coli*. Antibiotic treatment combining ultrasound with aminoblyciosides proved to have synergistic effect both for Gram-positive and Gram-negative bacteria [25]. Biofilms are highly tolerant to antibiotics and the host immune system, often associated with chronic disease states that create long-term problems for patients. Nishikawa et al. [26] studied the cavitation effect of ultrasound to develop an efficient plaque-control method for biofilms composed of *Streptococcus mutans*, which is often responsible for acid production and the progression of caries.

Although recent studies have suggested that the existence of UFBs alone impacts the morphology of certain bacteria and thickens the membranes, resulting in inhibited bacterial cell growth [27]. The first report using UFBs with antibiotics was from Ikeda-Dantsuji et al. [28], who demonstrated killing *Chlamydia trachomatis*-infected human epithelial cells. They succeeded in using ultrasound and liposome-based UFBs (average diameter 150 to 200 nm) to enhance the delivery of antibiotic molecules, in this case doxycycline and ceftizoxime, into cells to eradicate intracellular *Chlamydiae* bacteria without causing damage to the cells themselves. Similarly, a recent report showed prolonged drug release kinetics of perfluoropentane-cored UFBs, which were loaded with vancomycin, coupled to the outer dextran sulfate shell [29]. These novel UFBs were effective for methicillin-resistant *Staphylococcus aureus* infections in the skin and pharmaceutically stable for possible topical treatment of difficult-to-cure wound infections [30].

9.6 Oxygen Carriers

Warburg described that cancer cells maintain a high glycolytic rate even in conditions of adequate oxygen supply, which is known today as the "Warburg effect." Most of the malignant tumors are subjected to hypoxic conditions (low levels of oxygen) due to the disordered vasculature developed to supply oxygen to the rapidly growing tumor. Thus, glycolysis is essential for tumor survival and spread. Nevertheless, tumor hypoxia leads to a poor prognosis [31] due to a potential increase in malignancy and resistance among cancer patients to chemotherapy and radiation treatment. Hypoxia can also occur in the intercellular space in the tumor. There is a necessity to find a method to improve this condition for reducing tumor hypoxia in order to elevate the efficacy of current cancer therapies. In the recent years, much research has been devoted to evaluating the possibility of administration of oxygen-gas-filled UFBs to reverse the hypoxic environment in malignant tumors [32].

Owen et al. [33] reported oral administration of oxygen UFBs, which resulted in reducing mouse xenograft-tumor model for human pancreatic cancer. The administrated UFB diameter was relatively small, mostly less than 100 nm, but results showed significant tumor size change and a reduction in HIF1α expression both at transcriptional (mRNA) and translational (HIF1α protein) processes. Oxygenated water and Argon UFBs were administrated to mice as controls in this experiment. A recently published study demonstrated the effectiveness of doxorubicin-loaded oxygen UFBs to MDA-MB-231 breast cancer and HeLa cells in vitro [10]. The UFB size was in the range of 200 to 300 nm in this study. Researchers have fabricated different shell material UFBs for the same purpose. Oxygen-encapsulated nano-sized carboxymethyl cellulosic nanobubbles in the sub-100 nm size range were developed for mitigating the hypoxic regions of tumors to weaken the hypoxia-driven pathways and inhibit tumor growth. Hypomethylation of 5-methylcytosine in hypoxic regions of a tumor was reverted to enhance cancer treatment by epigenetic regulation, both in vitro and in vivo [34]. Others have developed UFBs with materials such as chitosan, phospholipids, polyethylene glycol and its various derivatives [10, 15, 35, 36]. Iijiima et al. [37] reported single nanometer-sized oxygen UFBs to overcome the hypoxia-induced resistance to radiation therapy via the suppression of hypoxia-inducible factor-1α. In addition, attempts to visualize and control the oxygen UFB within the cancer cells have produced interesting results in understanding the mechanism involved in this new therapeutic approach [38]. However, as most of the UFBs mentioned above are relatively small in size, more detailed evaluation is anticipated to confirm the existence of UFBs in in vivo studies.

9.7 Gene Therapy

Tachibana et al. [39, 40] were the first to report successful gene therapy in both in vitro and in vivo experiments by using ultrasound contrast agent MBs. Naked DNA nonviral gene vectors are more advantageous due to availability, cost-effectiveness, and less induction of immune system than viral vectors, which have made them an attractive candidate for gene delivery. However, their transfection efficiency is limited due to the rapid degradation by DNAse and non-tissue specificity. Microbubbles played a major role in increasing target specificity to various types of tissue and enhancing the gene transfection efficiency. This technology has recently been directly extended to a broader range of applications with UFBs for various gene therapies. Among the various applications of UFBs in medicine, this is probably the greatest advancement in the recent years. The large amount of literature on gene therapy with nano-sized bubbles shows the extent of interest and urgent demand to put this therapy to practice in patients.

We recently showed in our experiments the feasibility of fabrication of ultrasound-responsive albumin-based UFBs and optimized an in vitro experimental method to transfer genes into cancer cells. The new easy-to-use approach combines simultaneous ultrasound irradiation in a 96-well standard plate containing cultured cancer cells. A special film transparent to sonic waves played an important role in providing proper conditions for ultrasound exposure. The schematic representation of the preparation of the multi-well for sonication is shown in Fig. 9.3. Luciferase-expressing pDNA was introduced into HSC-2 cells using various solutions with or without NBs (UFBs) and sonication. Twenty-four hours after ultrasonic irradiation, the expression level of luciferase markedly increased in the existence of UFBs (Fig. 9.3). Although our UFBs were albumin shelled and targeted to cancer cells, in the recent few years, researchers have reported lipid-shelled UFBs for gene therapy in various organs and target diseases (Table 9.1).

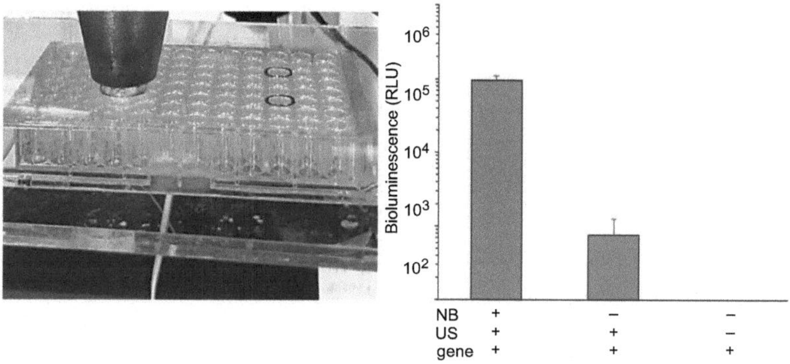

Figure 9.3 A 96-culture well plate ultrasound irradiation method. Luciferase-expressing pDNA was introduced into HSC2 cells with UFB/ultrasound.

Table 9.1 Literature on UFB-related gene therapy in various organs

Target Organ	Author/Year
Colon	R. Abdalkader/2017 [41]
Peritoneum	K. Nishimura/2017, 2019 [42, 43]
Liver	B. Wu/2016 [44], X. Xie/2016 [19], C. Huang/2017 [45], B. Zhang/2018 [46]
Brain	Y. Negishi/2015 [47], W. Cai/2018 [48], K. Ogawa/2019 [49], B. Cheng/2019 [50, 51]
Kidney/breast/ cervical	H. Jing/2016 [52], B. Tayier/2019 [12], Y. Peng/2019 [4]
Prostate	H. P. Tong/2013 [53], L. Wang/2014 [54], X. Fan/2016 [55], M. Wu/2018 [56]
Spinal cord	Z. Song/2017, 2018 [57, 58]
Skeletal muscles	Y. Watanabe/2010 [59]
Leukemia	E. Y. Lukianova-Hleb/2011 [60]
Retina	S. S. Thakur/2017 [61]

9.8 Ultrasound Imaging

Molecular targeted contrast-enhanced ultrasound has emerged as a very promising noninvasive imaging strategy. The advantages of ultrasound imaging, in general, are its high spatial and temporal resolution, obtainable with relatively low costs, portability, absence of ionizing irradiation, and of course the widespread availability of echograph equipment in many hospitals. Nevertheless, molecular ultrasound imaging techniques with MBs and UFBs are currently the major interest for many researchers and clinicians in this field. UFBs for ultrasound imaging have gained great momentum and are considered the next-generation molecular imaging modality.

In early studies on UFB ultrasound imaging, Wang et al. [62] fabricated the UFB contrast agents from soybean lipid, tween 80 surfactant, and cholesterol shell filled with sulfur hexafluoride gas. Shang et al. [63] also reported successful ultrasound contrast imaging in rabbits with octafluoropropane-filled UFBs, generated by span 60 and tween 80, resulting in relatively larger nano-sized bubbles ranging from 450 nm to 700 nm. A large amount of literature on acoustic UFB imaging has been reported as of now. Some UFBs have dual function ability for both imaging and therapy. The UFB shell materials have become more and more complex as therapeutic targets are highly specified to a certain disease or location in the body. Among the many reports, a particular ultrasound contrast agent of great interest is molecular targeted UFB for prostate cancer imagining. These UFBs are coupled with specific anti-PSMA (prostate specific membrane antigen) nano-bodies and are capable of binding to prostate tissues. These "targeted" UFBs are constructed via a biotin–streptavidin system, having an average diameter of about 500 nm. Anti-PSMA nanobody-carrying UFBs could specifically adhere to prostate cancer cells, thus resulting in improved diagnosis of malignant cancer. Ultrasonography indicators of UFB imaging (arrival time, peak time, peak intensity, and enhanced duration) were evaluated for three types of cancer in animal xenografts (LNCaP, C4-2, and MKN45). Ultrasound imaging observation showed that these three indicators of targeted UFBs exhibited significant differences from non-prostate cancer in animals [55]. UFBs could be combined with other therapeutic modalities as gene therapy or chemotherapy, such agents as androgen receptor siRNA or doxorubicin, resulting in the inhibition of prostate cancer growth [54]. Basically, the same molecular targeting mechanism could be applied for ovarian cancer with CA-125-targeted UFB contrast agents [64], TAG-72 antigen for breast and colon cancer [14].

9.9 High-Intensity Focused Ultrasound

Thermal ultrasound therapy for the brain and related neurological disorders has been investigated for over half a century. Although the therapeutic potential of ultrasound has long been recognized throughout the years, it was only until the late 1990s that breakthrough technology permitted clinicians to actually transmit ultrasound through the skull bone and into the brain. Adoption of phased array technology allowed ultrasound to focus on a sharp and localized area in the brain and thus to mechanically and thermally treat brain tissue. This can be carried out in a clinical situation in combination with magnetic resonance imaging (MRI) device in a noninvasive or minimally invasive manner. The novel technology has recently created a great potential to impact patient care for opening of the BBB, treating brain tumors, and Alzheimer's disease. This so-called high-intensity focused ultrasound (HIFU) therapy for the brain can be with no exception used in conjunction with MBs and UFBs for drug delivery. Current research on the application of UFBs for the brain mostly concentrates on ultrasound molecular imaging and drug/gene carriers. The research group lead by Exner and Chopra et al. [50, 51, 65] succeeded in the characterization of different bubble formulations for BBB opening, using a focused ultrasound system with acoustic feedback control, which includes UFBs. The experiments were conducted using a 0.5 MHz focused ultrasound transducer with in vivo focal pressure in the range of 0.1–0.7 MPa. Successful feedback control was achieved with UFB resulting in localized BBB opening confirmed with Evans blue dye leakage. Negishi et al. [47] demonstrated enhancement of BBB permeability and delivery of antisense oligonucleotides or plasmid DNA to the brain by the combination of lipid-based UFBs and HIFU. The importance of accumulating nucleic-acids-loaded UFBs at the BBB component cells was emphasized, to increase the delivery efficacy of antisense oligonucleotides or gene-expressing plasmid DNAs for brain parenchymal cells. There has been attempts to use low-intensity focused ultrasound for gene delivery for the acute spinal cord injury treatment. A novel type of brain-derived neurotrophic factor (BDNF)-loaded cationic UFBs conjugated with MAP-2 antibody gene therapy was applied with low-intensity focused ultrasound [58].

The HIFU technology can also be applied to various organs other than the brain. Huang et al. [45] recently reported synthesized PFP-encapsulated PAA–F127 thermosensitive UFBs with shell thicknesses of 20–50 nm, and ~5 nm SPIO Fe_3O_4 super-paramagnetic nanoparticles were loaded in the shells to significantly enhance the ultrasonic contrast and the magnetic susceptibility. The in vivo MR susceptibility experiments demonstrated that the SPIO-embedded nanobubbles (UFBs) can be magnetically guided to the tumor to locally enhance the contrast in ultrasound and MR imaging. Furthermore, this UFB enhanced the passive and physical targeting of tumors resulting in HIFU-triggered drug release for tumor therapy. HIFU has also been used in combination with UFBs for breast and colon cancers in animals [8].

9.10 Limitations

In order to actually have UFBs to become a Food and Drug Administration (FDA)-approved diagnostic or therapeutic agent requires rigorous clinical trials to show safety and efficacy. A randomized trial design is the only reliable means to determine the treatment effect and diagnostic accuracy. A strict safety standard, quality-controlled monitoring system must be securely constructed as a product. These issues have to be addressed in a straightforward manner. Although there are so much experimental data coming from multiple labs around the world in such short durations, the technical question we first encounter is how accurately can we characterize UFBs? It could be assumed that this question is not easy to answer. Contamination of nanoparticles could interfere with the therapeutic outcome in some cases. Some researchers have boldly stood up to challenge this issue. Hernandez et al. [66] evaluated the use of resonant mass measurement technique to separate nanoparticles and nanobubbles. In order to ensure the quality of UFBs, such measurements must be simultaneously conducted with multiple modalities. We recently reported UFB measurement data with real-time acoustic characterization as well as with several kinds of conventional methods [11, 67]. In spite of recent progress, it is still safe to say that the measurement of UFBs with an estimated diameter under 100 nm still remains a challenge in most cases, especially for medically applicable UFBs.

9.11 Conclusion

Encapsulated UFBs in medicine are still an emerging technology but are very promising as a future theranostic agent. The technology has a potential to become a major player in noninvasive ultrasonic diagnosis and therapy for a broad range of diseases.

References

1. Frinking, P., Segers, T., Luan, Y., and Tranquart, F. Three decades of ultrasound contrast agents: A review of the past, present and future improvements. *Ultrasound Med. Biol.* 2020; **46**(4): 892–908.
2. Tachibana, K. and Tachibana, S. Albumin microbubble echo-contrast material as an enhancer for ultrasound accelerated thrombolysis. *Circulation* 1995; **92**(5): 1148–1150.
3. Tachibana, K., Uchida, T., Ogawa, K., Yamashita, N., and Tamura, K. Induction of cell-membrane porosity by ultrasound. *Lancet* 1999; **353**(9162).
4. Peng, Y., Zhu, L., Wang, L., Liu, Y., Fang, K., Lan, M., et al. Preparation of nanobubbles modified with a small-molecule CXCR4 antagonist for targeted drug delivery to tumors and enhanced ultrasound molecular imaging. *Int. J. Nanomed.* 2019; **14**: 9139–9157.
5. Abenojar, E. C., Nittayacharn, P., de Leon, A. C., Perera, R., Wang, Y., Bederman, I., et al. Effect of bubble concentration on the in vitro and in vivo performance of highly stable lipid shell-stabilized micro- and nanoscale ultrasound contrast agents. *Langmuir* 2019; **35**(31): 10192–10202.
6. Bosca, F., Bielecki, P. A., Exner, A. A., and Barge, A. Porphyrin-loaded pluronic nanobubbles: A new US-activated agent for future theranostic applications. *Bioconjug. Chem.* 2018; **29**(2): 234–240.

7. Chandan, R. and Banerjee, R. Pro-apoptotic liposomes-nanobubble conjugate synergistic with paclitaxel: A platform for ultrasound responsive image-guided drug delivery. *Sci. Rep.* 2018; **8**(1): 2624.
8. Wu, H., Abenojar, E. C., Perera, R., de Leon, A. C., An, T., and Exner, A. A. Time-intensity-curve analysis and tumor extravasation of nanobubble ultrasound contrast agents. *Ultrasound Med. Biol.* 2019; **45**(9): 2502–2514.
9. Kida, H., Nishimura, K., Ogawa, K., Watanabe, A., Feril, L. B., Irie, Y., et al. Nanobubble mediated gene delivery in conjunction with a hand-held ultrasound scanner. *Front. Pharmacol.* 2020; **11**: 363.
10. Khan, M. S., Hwang, J., Lee, K., Choi, Y., Jang, J., Kwon, Y., et al. Surface composition and preparation method for oxygen nanobubbles for drug delivery and ultrasound imaging applications. *Nanomaterials* (Basel) 2019; **9**(1).
11. Lafond, M., Watanabe, A., Yoshizawa, S., Umemura, S. I., and Tachibana, K. Cavitation-threshold determination and rheological-parameters estimation of albumin-stabilized nanobubbles. *Sci. Rep.* 2018; **8**(1): 7472.
12. Tayier, B., Deng, Z., Wang, Y., Wang, W., Mu, Y., and Yan, F. Biosynthetic nanobubbles for targeted gene delivery by focused ultrasound. *Nanoscale* 2019; **11**(31): 14757–14768.
13. Xu, J. S., Huang, J., Qin, R., Hinkle, G. H., Povoski, S. P., Martin, E. W., et al. Synthesizing and binding dual-mode poly (lactic-co-glycolic acid) (PLGA) nanobubbles for cancer targeting and imaging. *Biomaterials* 2010; **31**(7): 1716–1722.
14. Zhang, X., Zheng, Y., Wang, Z., Huang, S., Chen, Y., Jiang, W., et al. Methotrexate-loaded PLGA nanobubbles for ultrasound imaging and synergistic targeted therapy of residual tumor during HIFU ablation. *Biomaterials* 2014; **35**(19): 5148–5161.
15. Khan, M. S., Hwang, J., Lee, K., Choi, Y., Seo, Y., Jeon, H., et al. Anti-tumor drug-loaded oxygen nanobubbles for the degradation of HIF-1alpha and the upregulation of reactive oxygen species in tumor cells. *Cancers* (Basel) 2019; **11**(10).
16. Lin, W., Xie, X., Deng, J., Liu, H., Chen, Y., Fu, X., et al. Cell-penetrating peptide-doxorubicin conjugate loaded NGR-modified nanobubbles for ultrasound triggered drug delivery. *J. Drug Target.* 2016; **24**(2): 134–146.
17. Shen, S., Li, Y., Xiao, Y., Zhao, Z., Zhang, C., Wang, J., et al. Folate-conjugated nanobubbles selectively target and kill cancer cells via

ultrasound-triggered intracellular explosion. *Biomaterials* 2018; 181: 293–306.

18. Tian, Y., Liu, Z., Zhang, L., Zhang, J., Han, X., Wang, Q., et al. Apatinib-loaded lipid nanobubbles combined with ultrasound-targeted nanobubble destruction for synergistic treatment of HepG2 cells in vitro. *Onco Targets Ther.* 2018; **11**: 4785–4795.
19. Xie, X., Lin, W., Liu, H., Deng, J., Chen, Y., Liu, H., et al. Ultrasound-responsive nanobubbles contained with peptide-camptothecin conjugates for targeted drug delivery. *Drug Deliv.* 2016; **23**(8): 2756–2764.
20. Liu, J., Chen, Y., Wang, G., Lv, Q., Yang, Y., Wang, J., et al. Ultrasound molecular imaging of acute cardiac transplantation rejection using nanobubbles targeted to T lymphocytes. *Biomaterials* 2018; **162**: 200–207.
21. Xie, F., Li, Z. P., Wang, H. W., Fei, X., Jiao, Z. Y., Tang, W. B., et al. Evaluation of liver ischemia-reperfusion injury in rabbits using a nanoscale ultrasound contrast agent targeting ICAM-1. *PLoS One* 2016; **11**(4): e0153805.
22. Kun-kun, W., Bin-bin, G., Feng, Z., Xiao-wei, W., Jiemei, Y., Ya-ru, Z., and Chang-jun, W. Design of ligands-conjuncted lipid nanobubbles as ultrasound contrast agents targeted to atherosclerotic plaques. *J. Nanosci. Nanotechnol.* 2016; **16**(7): 7611–7616.
23. Sutton, J. T., Haworth, K. J., Pyne-Geithman, G., and Holland, C. K. Ultrasound-mediated drug delivery for cardiovascular disease. *Expert Opin. Drug Deliv.* 2013; **10**(5): 573–592.
24. Li, M., Liu, Y., Chen, J., Liu, T., Gu, Z., Zhang, J., et al. Platelet bio-nanobubbles as microvascular recanalization nanoformulation for acute ischemic stroke lesion theranostics. *Theranostics* 2018; **8**(18): 4870–4883.
25. Pitt, W. G., McBride, M. O., Lunceford, J. K., Roper, R. J., and Sagers, R. D. Ultrasonic enhancement of antibiotic action on gram-negative bacteria. *Antimicrob. Agents Chemother.* 1994; **38**(11): 2577–2582.
26. Nishikawa, T., Yoshida, A., Khanal, A., Habu, M., Yoshioka, I., Toyoshima, K., et al. A study of the efficacy of ultrasonic waves in removing biofilms. *Gerodontology* 2010; **27**(3): 199–206.
27. Luu, T. Q., Hong Truong, P. N., Zitzmann, K., and Nguyen, K. T. Effects of ultrafine bubbles on gram-negative bacteria: Inhibition or selection? *Langmuir* 2019; **35**(42): 13761–13768.

28. Ikeda-Dantsuji, Y., Feril, L. B., Jr., Tachibana, K., Ogawa, K., Endo, H., Harada, Y., et al. Synergistic effect of ultrasound and antibiotics against *Chlamydia trachomatis*-infected human epithelial cells in vitro. *Ultrason. Sonochem.* 2011; **18**(1): 425–430.

29. Argenziano, M., Banche, G., Luganini, A., Finesso, N., Allizond, V., Gulino, G. R., et al. Vancomycin-loaded nanobubbles: A new platform for controlled antibiotic delivery against methicillin-resistant *Staphylococcus aureus* infections. *Int. J. Pharm.* 2017; **523**(1): 176–188.

30. Ronan, E., Edjiu, N., Kroukamp, O., Wolfaardt, G., and Karshafian, R. USMB-induced synergistic enhancement of aminoglycoside antibiotics in biofilms. *Ultrasonics* 2016; **69**: 182–190.

31. Bhandari, P. N., Cui, Y., Elzey, B. D., Goergen, C. J., Long, C. M., and Irudayaraj, J. Oxygen nanobubbles revert hypoxia by methylation programming. *Sci. Rep.* 2017; **7**(1): 9268.

32. Orel, V. B., Zabolotny, M. A., and Orel, V. E. Heterogeneity of hypoxia in solid tumours and mechanochemical reactions with oxygen nanobubbles. *Med. Hypotheses* 2017; **102**: 82–86.

33. Owen, J., McEwan, C., Nesbitt, H., Bovornchutichai, P., Averre, R., Borden, M., et al. Reducing tumour hypoxia via oral administration of oxygen nanobubbles. *PLoS One* 2016; **11**(12): e0168088.

34. Bhandari, P., Novikova, G., Goergen, C. J., and Irudayaraj, J. Ultrasound beam steering of oxygen nanobubbles for enhanced bladder cancer therapy. *Sci. Rep.* 2018; **8**(1): 3112.

35. Cavalli, R., Bisazza, A., Rolfo, A., Balbis, S., Madonnaripa, D., Caniggia, I., et al. Ultrasound-mediated oxygen delivery from chitosan nanobubbles. *Int. J. Pharm.* 2009; **378**(1–2): 215–217.

36. Khan, M. S., Hwang, J., Lee, K., Choi, Y., Kim, K., Koo, H. J., et al. Oxygen-carrying micro/nanobubbles: Composition, synthesis techniques and potential prospects in photo-triggered theranostics. *Molecules* 2018; **23**(9).

37. Iijima, M., Gombodorj, N., Tachibana, Y., Tachibana, K., Yokobori, T., Honma, K., et al. Development of single nanometer-sized ultrafine oxygen bubbles to overcome the hypoxia-induced resistance to radiation therapy via the suppression of hypoxia-inducible factor1alpha. *Int. J. Oncol.* 2018; **52**(3): 679–686.

38. Bhandari, P., Wang, X., and Irudayaraj, J. Oxygen nanobubble tracking by light scattering in single cells and tissues. *ACS Nano* 2017; **11**(3): 2682–2688.

39. Li, T., Tachibana, K., and Kuroki, M. Gene transfer with echo-enhanced contrast agents: Comparison between Albunex, Optison, and Levovist in mice—initial results. *Radiology* 2003; **229**(2): 423–428.

40. Taniyama, Y., Tachibana, K., Hiraoka, K., Namba, T., Yamasaki, K., Hashiya, N., et al. Local delivery of plasmid DNA into rat carotid artery using ultrasound. *Circulation* 2002; **105**(10): 1233–1239.

41. Abdalkader, R., Kawakami, S., Unga, J., Higuchi, Y., Suzuki, R., Maruyama, K., et al. The development of mechanically formed stable nanobubbles intended for sonoporation-mediated gene transfection. *Drug Deliv.* 2017; **24**(1): 320–327.

42. Nishimura, K., Fumoto, S., Fuchigami, Y., Hagimori, M., Maruyama, K., and Kawakami, S. Effective intraperitoneal gene transfection system using nanobubbles and ultrasound irradiation. *Drug Deliv.* 2017; **24**(1): 737–744.

43. Nishimura, K., Yonezawa, K., Fumoto, S., Miura, Y., Hagimori, M., Nishida, K., et al. Application of direct sonoporation from a defined surface area of the peritoneum: Evaluation of transfection characteristics in mice. *Pharmaceutics* 2019; **11**(5).

44. Wu, B., Qiao, Q., Han, X., Jing, H., Zhang, H., Liang, H., et al. Targeted nanobubbles in low-frequency ultrasound-mediated gene transfection and growth inhibition of hepatocellular carcinoma cells. *Tumour Biol.* 2016; **37**(9): 12113–12121.

45. Huang, C., Zhang, H., and Bai, R. Advances in ultrasound-targeted microbubble-mediated gene therapy for liver fibrosis. *Acta Pharm. Sin. B* 2017; **7**(4): 447–452.

46. Zhang, B., Chen, M., Zhang, Y., Chen, W., Zhang, L., and Chen, L. An ultrasonic nanobubble-mediated PNP/fludarabine suicide gene system: A new approach for the treatment of hepatocellular carcinoma. *PLoS One* 2018; **13**(5): e0196686.

47. Negishi, Y., Yamane, M., Kurihara, N., Endo-Takahashi, Y., Sashida, S., Takagi, N., et al. Enhancement of blood–brain barrier permeability and delivery of antisense oligonucleotides or plasmid DNA to the brain by the combination of bubble liposomes and high-intensity focused ultrasound. *Pharmaceutics* 2015; **7**(3): 344–362.

48. Cai, W., Lv, W., Feng, Y., Yang, H., Zhang, Y., Yang, G., et al. The therapeutic effect in gliomas of nanobubbles carrying siRNA combined with ultrasound-targeted destruction. *Int. J. Nanomed.* 2018; **13**: 6791–6807.

49. Ogawa, K., Fuchigami, Y., Hagimori, M., Fumoto, S., Maruyama, K., and Kawakami, S. Ultrasound-responsive nanobubble-mediated gene transfection in the cerebroventricular region by intracerebroventricular administration in mice. *Eur. J. Pharm. Biopharm.* 2019; **137**: 1–8.

50. Cheng, B., Bing, C., and Chopra, R. The effect of transcranial focused ultrasound target location on the acoustic feedback control performance during blood-brain barrier opening with nanobubbles. *Sci. Rep.* 2019; **9**(1): 20020.

51. Cheng, B., Bing, C., Xi, Y., Shah, B., Exner, A. A., and Chopra, R. Influence of nanobubble concentration on blood–brain barrier opening using focused ultrasound under real-time acoustic feedback control. *Ultrasound Med. Biol.* 2019; **45**(8): 2174–2187.

52. Jing, H., Cheng, W., Li, S., Wu, B., Leng, X., Xu, S., et al. Novel cell-penetrating peptide-loaded nanobubbles synergized with ultrasound irradiation enhance EGFR siRNA delivery for triple negative breast cancer therapy. *Colloid. Surf. B: Biointerfaces* 2016; **146**: 387–395.

53. Tong, H. P., Wang, L. F., Guo, Y. L., Li, L., Fan, X. Z., Ding, J., et al. Preparation of protamine cationic nanobubbles and experimental study of their physical properties and in vivo contrast enhancement. *Ultrasound Med. Biol.* 2013; **39**(11): 2147–2157.

54. Wang, L., Zhang, M., Tan, K., Guo, Y., Tong, H., Fan, X., et al. Preparation of nanobubbles carrying androgen receptor siRNA and their inhibitory effects on androgen-independent prostate cancer when combined with ultrasonic irradiation. *PLoS One* 2014; **9**(5): e96586.

55. Fan, X., Guo, Y., Wang, L., Xiong, X., Zhu, L., and Fang, K. Diagnosis of prostate cancer using anti-PSMA aptamer A10-3.2-oriented lipid nanobubbles. *Int. J. Nanomed.* 2016; **11**: 3939–3950.

56. Wu, M., Zhao, H., Guo, L., Wang, Y., Song, J., Zhao, X., et al. Ultrasound-mediated nanobubble destruction (UMND) facilitates the delivery of A10-3.2 aptamer targeted and siRNA-loaded cationic nanobubbles for therapy of prostate cancer. *Drug Deliv.* 2018; **25**(1): 226–240.

57. Song, Z., Wang, Z., Shen, J., Xu, S., and Hu, Z. Nerve growth factor delivery by ultrasound-mediated nanobubble destruction as a treatment for acute spinal cord injury in rats. *Int. J. Nanomed.* 2017; **12**: 1717–1729.

58. Song, Z., Ye, Y., Zhang, Z., Shen, J., Hu, Z., Wang, Z., et al. Noninvasive, targeted gene therapy for acute spinal cord injury using LIFU-mediated BDNF-loaded cationic nanobubble destruction. *Biochem. Biophys. Res. Commun.* 2018; **496**(3): 911–920.

59. Watanabe, Y., Horie, S., Funaki, Y., Kikuchi, Y., Yamazaki, H., Ishii, K., et al. Delivery of Na/I symporter gene into skeletal muscle using nanobubbles and ultrasound: visualization of gene expression by PET. *J. Nucl. Med.* 2010; **51**(6): 951–958.

60. Lukianova-Hleb, E. Y., Samaniego, A. P., Wen, J., Metelitsa, L. S., Chang, C. C., and Lapotko, D. O. Selective gene transfection of individual cells in vitro with plasmonic nanobubbles. *J. Control. Release* 2011; **152**(2): 286–293.

61. Thakur, S. S., Ward, M. S., Popat, A., Flemming, N. B., Parat, M. O., Barnett, N. L., et al. Stably engineered nanobubbles and ultrasound: An effective platform for enhanced macromolecular delivery to representative cells of the retina. *PLoS One* 2017; **12**(5): e0178305.

62. Wang, Y., Li, X., Zhou, Y., Huang, P., and Xu, Y. Preparation of nanobubbles for ultrasound imaging and intracelluar drug delivery. *Int. J. Pharm.* 2010; **384**(1–2): 148–153.

63. Shang, M., Wang, K., Guo, L., Duan, S., Lu, Z., and Li, J. Development of novel ST68/PLA-PEG stabilized ultrasound nanobubbles for potential tumor imaging and theranostic. *Ultrasonics* 2019; **99**: 105947.

64. Gao, Y., Hernandez, C., Yuan, H. X., Lilly, J., Kota, P., Zhou, H., et al. Ultrasound molecular imaging of ovarian cancer with CA-125 targeted nanobubble contrast agents. *Nanomedicine* 2017; **13**(7): 2159–2168.

65. Bing, C., Hong, Y., Hernandez, C., Rich, M., Cheng, B., Munaweera, I., et al. Characterization of different bubble formulations for blood-brain barrier opening using a focused ultrasound system with acoustic feedback control. *Sci. Rep.* 2018; **8**(1): 7986.

66. Hernandez, C., Abenojar, E. C., Hadley, J., de Leon, A. C., Coyne, R., Perera, R., et al. Sink or float? Characterization of shell-stabilized bulk nanobubbles using a resonant mass measurement technique. *Nanoscale* 2019; **11**(3): 851–855.

67. Watanabe, A., Sheng, H., Endo, H., Feril, L. B., Irie, Y., Ogawa, K., et al. Echographic and physical characterization of albumin-stabilized nanobubbles. *Heliyon* 2019; **5**(6): e01907.

Chapter 10

Dental Application of Ozone Ultrafine Bubble Water

Shinichi Arakawa
Department of Lifetime Oral Health Care Science, Graduate School, Tokyo Medical and Dental University, 1-5-45 Yushima, Bunkyo-ku, Tokyo 113-8549, Japan
shinperi@tmd.ac.jp

Ozone has strong antimicrobial effects and does not induce microbial resistance. Aqueous ozone possesses a high level of biocompatibility. However, ozonated water has a half-life of about only 20 min and will change to oxygen very quickly. To overcome this instability of ozonated water, ozone ultrafine bubble water (OUFBW) has been developed. It has been revealed that OUFBW is useful and effective in therapy for periodontitis and peri-implantitis. Furthermore, OUFBW might have biological activity in addition to antimicrobial effect. OUFBW induces the production of reactive oxygen species (ROS) in an ozone-dependent manner. The oxidative stress caused by OUFBW activates the mitogen-activated protein kinase (MAPK) pathway, especially p38 MAPK, and promotes the translocation of both c-Fos and nuclear factor erythroid 2 [NF-E2]-related factor 2

Ultrafine Bubbles
Edited by Koichi Terasaka, Kyuichi Yasui, Wataru Kanematsu, and Nobuhiro Aya

ISBN 978-981-4877-59-6 (Hardcover), 978-1-003-14195-2 (eBook)
www.jennystanford.com

(Nrf2) into the nuclei. The transcription factor c-Fos is involved in the mechanism of osteoblastic cells differentiation, and Nrf2 can bind with antioxidant response element (ARE) to induce cytoprotective responses to oxidative stress. OUFBW accelerates wound healing via tolerance induction against oxidase stresses rather than via antibacterial effects.

10.1 Periodontal Therapy

10.1.1 Periodontitis

Periodontitis is a chronic inflammatory disease caused by microorganisms existing in subgingival pockets. *Porphyromonas gingivalis, Treponema denticola*, and *Tannerella forsythia* are the specific pathogens most frequently associated with this disease, being present in high numbers within deep periodontal pockets in severe periodontal lesions [1]. Elimination of pathogens containing biofilms is the primary goal of periodontal treatment. Supra- and subgingival mechanical debridement is traditionally used as the initial phase of treatment to achieve this goal even though it is rarely capable of complete removal of periodontal pathogens. Petersilka et al. reported that the effectiveness of subgingival debridement procedures was poor, and deposits such as plaque or tartar are difficult to be removed especially in deep pockets over 4 mm depths or root furcation area [2].

10.1.2 Periodontal Treatment

The most effective elimination of biofilm at periodontal lesions is mechanical debridement by using toothbrush, interdental brush, and dental floss for each patient and by utilizing hand scalers or ultrasonic scaler in periodontal therapy. On the other hand, in order to compensate for mechanical debridement, adjunctive antimicrobial agents in the form of topical or systemic antibiotics have been employed [3]. Furthermore, subgingival irrigation with various antiseptic agents such as povidone-iodine, chlorhexidine (CHX), or hydrogen peroxide has been performed in conjunction with scaling and root planing, which provided significant clinical benefits relative to conventional mechanical root debridement alone. However, these existing antiseptic agents have serious insufficiency, e.g., the prolonged use of CHX may cause mucosal desquamation, tooth staining, and altered taste sensation. In addition, an increasing number of immediate-type allergies to this agent such as anaphylactic shock has been reported [4]. Therefore, an alternative adjunctive antiseptic with high antimicrobial potential, a good safety profile, fewer contraindications, and a higher degree of usability would be beneficial for periodontal therapy.

10.1.3 Ozone Treatment

Ozone has strong antimicrobial effects against bacteria, fungi, protozoa, and viruses [5] and does not induce microbial resistance [6]. Aqueous ozone possesses a high level of biocompatibility to fibroblasts, cementoblasts, and epithelial cells [7–10]. So far, positive clinical effects of ozonated water in reducing signs of gingivitis and periodontitis have been reported in clinical studies [29, 30]. However, ozonated water has a half-life of about only 20 min and will change to oxygen very quickly.

To overcome this instability of ozonated water, ozone ultrafine bubble water (OUFBW; formerly ozone nanobubble water: NBW3) has been developed using nanobubble generating technology. Ozone is preserved for more than 6 months (data not shown). The ozone concentration of OUFBW is 1.5 mg/L, which is equivalent to the oxidation titer determined by electron spin resonance. Although the entire mechanisms of OUFBW to inactivate bacteria are not well known yet, it might be basically similar to those of the existing ozonated water. Ozone, per se, also changes to oxygen when it reacts with organic substances. In this process, hydroxyl radicals (·OH) are generated, which are the most reactive oxidizing species. These free radicals might play a role in killing bacteria effectively.

Previously, we confirmed that OUFBW had strong bactericidal effects against multiple-drug-resistant bacteria and representative periodontopathic bacteria in vitro [11. Furthermore, we confirmed that OUFBW was biocompatible with human oral tissues such as oral epithelium and mucosa in vitro [11]. It is authorized to be used as drinking water by the Ministry of Health, Labour and Welfare in Japan (No. 10- S1-2393), which indicates its high level of safety for use in humans. Due to those, we evaluated the clinical and microbiological effects of OUFBW irrigation as an adjunct to full-mouth ultrasonic subgingival debridement in the treatment of periodontitis.

10.1.4 OUFBW in Periodontal Treatment

Twenty-two subjects were randomly assigned to one of the two treatment groups: full-mouth mechanical debridement with tap water (WATER) or full-mouth mechanical debridement with OUFBW (NBW3). Clinical examination was performed at baseline and 4 and 8 weeks after treatment. Microbiological examination was carried out just before and after treatment and at 1 and 8 weeks post-treatment. There were significant improvements in all clinical parameters after 4 weeks in both groups. The reduction in the probing pocket depths after 4 and 8 weeks in the OUFBW group was significantly greater than that in the WATER group (Fig. 10.1). Moreover, the mean total number of bacteria in subgingival plaque over the study period was significantly reduced only in the OUFBW group at 1 and 8 weeks (Fig. 10.2). These results suggest that subgingival irrigation with OUFBW may be a valuable adjunct to periodontal treatment [12].

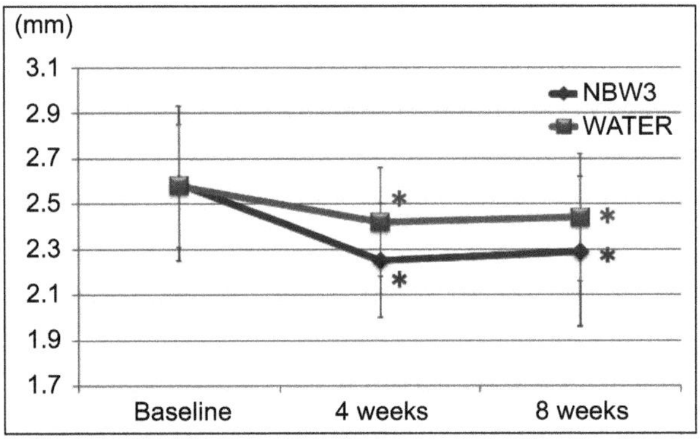

Figure 10.1 Changes in the full-mouth mean (± standard deviation) PPD (millimeter) (in each treatment group from baseline to 8 weeks). NBW3 (OUFBW) full-mouth debridement with ozone nanobubble water, WATER full-mouth debridement with tap water, BOP bleeding on probing, PPD probing pocket depth. No significant differences in all the clinical parameters assessed were seen between groups at any examination period ($p > 0.05$). *$p < 0.05$, within the groups, PPD decreased significantly from baseline to re-examination periods (paired t-test).

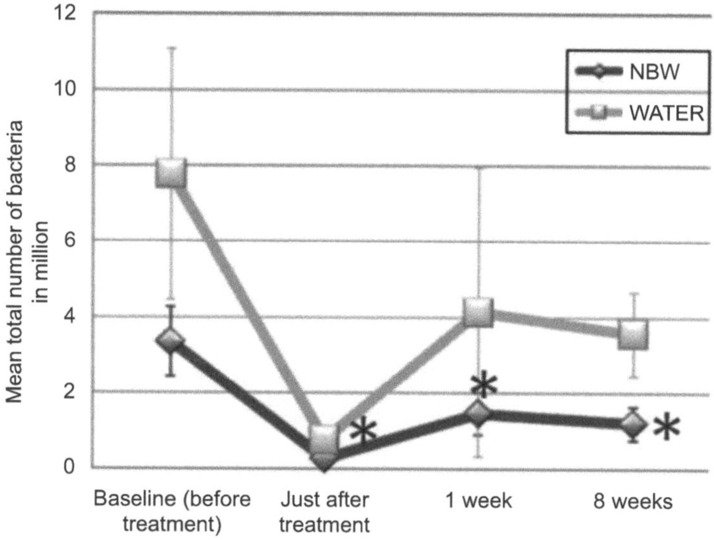

Figure 10.2 Changes in the mean (± standard deviation) total number of bacteria in subgingival plaque in each treatment group from baseline (before treatment) to 8 weeks. NBW3 (OUFBW) full-mouth debridement with ozone nanobubble water, WATER full-mouth debridement with water. No significant differences were seen between groups at any examination period ($p > 0.05$). *$p < 0.05$, within the groups, mean total number of bacteria in subgingival plaque decreased significantly from baseline (before treatment) to re-examination periods (paired t-test).

10.2 Therapy for Peri-implantitis

We investigated the effects of ONBW on peri-implantitis lesions with nonsurgical treatment [13]. The advanced quick bonding (AQB) implants on 19 and 20 were placed on a 43-year-old female in 2011. Her chief complaint was swelling at the site of 20 in 2015. The patient presented swelling at the peri-implant area, bleeding of probing (BOP), and a probing depth (PD) of 6 mm at the buccal site of an implant of 20.

There were no findings of mobility at the concerned implant. Bone resorption and a radiolucent part around the implant were confirmed with periapical radiograph (Fig. 10.3a). The amount of bone loss was 5.0 and 6.5 mm at mesial and distal sites of 20, respectively. This peri-implantitis was evaluated; the case required surgical therapy according to the cumulative interceptive supportive therapy (CIST). In addition to mechanical plaque control, the patient received regular professional oral hygiene treatment and irrigation with ONBW every week at 100 mL each. Also for chemical plaque control, the irrigation of pockets was performed at home by the patient at a frequency of three times a day every day. After 12 weeks, soft tissues of the peri-implant presented no clinical signs of inflammation and BOP, and PD was 3 mm (Fig. 10.3b). The bone levels did not change significantly as demonstrated by the follow-up roentgenography taken after 3 years (Fig. 10.3c). Microbiologically, the number of red complex of periodontopathic bacteria decreased substantially (data not shown).

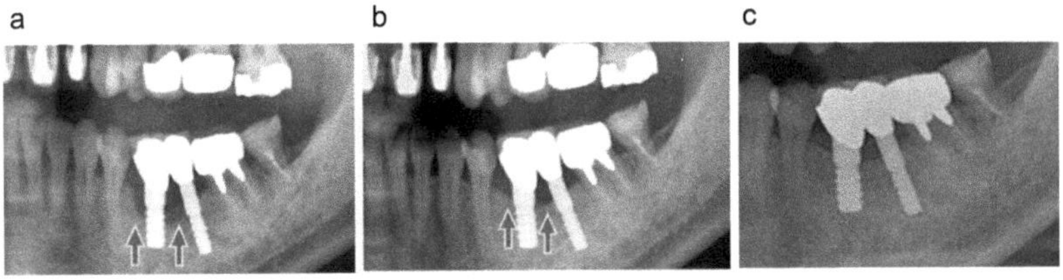

Figure 10.3 The amount of bone loss was 5.0 and 6.5 mm at mesial and distal sites of 20, respectively, at 0 week (a). The mesial and distal levels of the peri-implant marginal bone were recovered radiographically at 12 weeks (b) and this condition was stable following 3 years (c).

Irrigation with OUFBW has been proven to be an effective additional treatment with mechanical plaque control. In 2004, Lang et al. presented valuable guidelines based on a diagnostic therapeutic algorithm of CIST [14]. A surgical approach in addition to the implant surface decontamination and antibiotic therapy should be performed for peri-implantitis treatment. However, in this case bone regeneration around the implant was obtained by using OUFBW.

Gloria et al. reported that heterogeneous bacteria, including periodontopathic bacteria, uncultivable asaccharolytic anaerobic Gram-positive rods, and other uncultivable Gram-negative rods and opportunistic microorganisms, were infected in peri-implantitis lesions [15]. Erdemci et al. reported on the investigation of the effects of systemic and topical ozone utilization in alveolar bone regeneration after tooth extraction. They concluded that postoperative long-term systemic ozone application can accelerate alveolar bone healing following extraction [16]. Ozdemir et al. evaluated the effect of ozone therapy on autogenous bone graft in calvarial defects [17]. They concluded that ozone therapy provided to new bone formation with an autogenous bone graft in the rat calvarial defect model. Another study also described that ozone therapy was effective on bone formation in calvarial defects of rats [18]. The high stability of OUFBW allows for bottling and use as a disinfectant solution at both clinical room and patient's home easily. Future studies should compare the irrigation of ozone-inactivated OUFBW or other disinfectants and investigate the effects of UFBW to peri-implantitis related to the other kind of implants such as titanium or zirconium implant.

10.3 Future Prospects

10.3.1 Induction of Cellular Signaling Involved in Oxidative Stress Responses in Human Periodontal Ligament Fibroblasts

Since OUFBW is effective to periodontitis and peri-implantitis, we think that OUFBW might have biological activity in addition to antimicrobial effect. In this section, we discuss OUFBW-induced oxidative stress in cells, mediated by the production of ROS, and the oxidative stress induced activation of the mitogen-activated protein kinase (MAPK) pathway in the cells. OUFBW also triggered the activation of c-Fos, a major component of the transcription factor activator protein 1 (AP-1), and also nuclear factor erythroid 2 [NF-E2]-related factor 2 (Nrf2) transcription factor with a sensitivity to oxidative stress [19].

We used commercially available OUFBW (Kyocera Corp); a concentration of 2.5 ppm [measured by ozone meter (AOM-05)] was used in this study. The particle concentration of OUFBW was 1.68×10^9 particles/mL [determined by nanoparticle multi-analyzer: qNano (Meiwafosis)] (Fig. 10.4). Inactivation of ozone was performed by ultraviolet ray irradiation for 4 h. Human primary periodontal ligament fibroblasts (hPDLFs) isolated from a 16-year-old male was used for target cells.

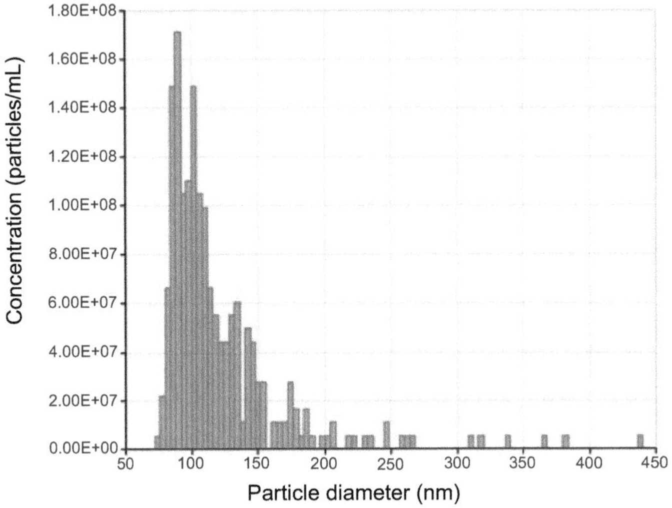

Particle diameter (nm)		**Concentration (particles/mL)**
Mean:	122 (Std Dev = 49.0)	Raw concentration: 1.68×10^9
Mode:	91	
Minimum:	76	
Maximum:	440	

Figure 10.4 Particle concentration of OUFBW.

The hPDLFs were exposed to OUFBW for 1 and 10 min because the treatment time with OUFBW is assumed to be less than 10 min in clinical use. The results showed that there were no significant differences in the cell viability between OUFBW-stimulated groups and negative control groups.

Next, we confirmed that induction of intracellular ROS depends on the ozone in OUFBW, mediated by oxidative stress. The phosphorylation of p38 MAPK was induced after treatment with OUFBW in hPDLFs, but not in those exposed to inactive OUFBW, indicating that p38 MAPK activation was ozone dependent. Furthermore, in OUFBW-treated hPDLFs, c-Fos and Nrf2 was translocated into their nuclei (Fig. 10.5).

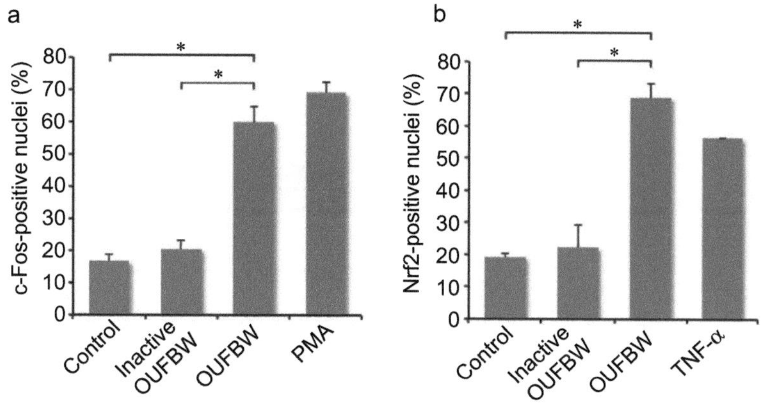

Figure 10.5 Nuclear translocation of c-Fos and Nrf2 induced by OUFBW. The hPDLFs were stimulated with the medium, inactive OUFBW or OUFBW, for 10 min. After the stimulation, the cells were incubated with the medium for 30 min or 6 h for evaluating the localization of c-Fos or Nrf2, respectively. Cells stimulated with 1 mM Phorbol Myristate Acetate (PMA) for 30 min or 500 ng/mL tumor necrosis factor-α (TNF-α) for 4 h were used as the positive controls. Scale bars = 100 μm. (c and d) Percentage of the cells showing nuclear translocation of c-Fos (a) or Nrf2 (b). **p* < 0.01 (Student's t-test). Data are shown as means ± SD.

RNA sequencing (RNA-seq) by using Illumina HiSeq platform revealed that 88 genes were significantly upregulated, while 80 genes were downregulated. The most upregulated gene is metallothionein-1G (*MT1G*, 59.5-fold), followed by HES1 (hairy and enhancer of split 1: *HES1*, 13.9-fold) and c-Fos (*FOS*, 10.2-fold). GO (www.geneontology.org) enrichment showed that the molecular function category was high in "binding," the cellular component category was concentrated in "cell part," and the biological process category was increased in "biological regulation." Top 30 pathways, including "mineral absorption," "neuroactive," and "MAPK signaling pathway," are high clustered and significantly enriched (q-value < 0.05) in KEGG enrichment analysis.

OUFBW induced the production of ROS in an ozone-dependent manner. The oxidative stress caused by OUFBW-induced activation of the MAPK pathway, especially p38 MAPK, induced the translocation of both c-Fos and Nrf2 into the nuclei (Fig. 10.6). The transcription factor c-Fos is involved in the mechanism of osteoblastic cells differentiation [20].

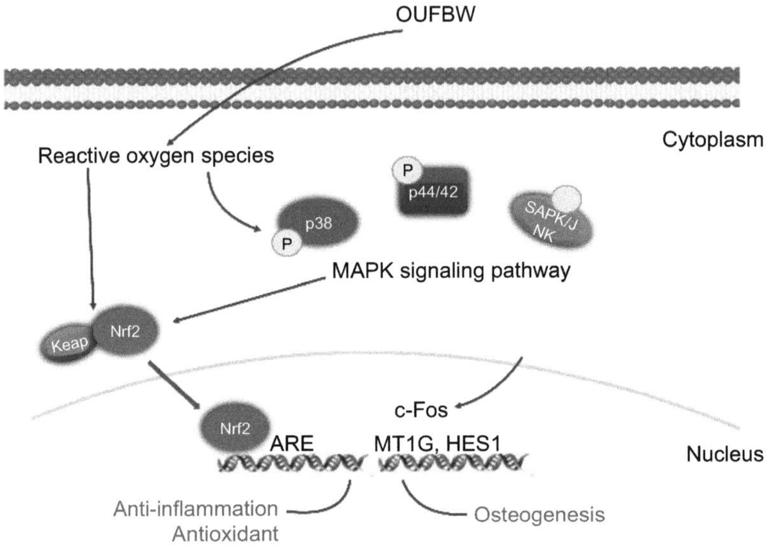

Figure 10.6 Summary of OUFBW function.

Therefore, we speculate that the downstream genes under the control of these transcription factors may play a role in regenerating periodontal tissues. In RNA-seq analysis, the most upregulated gene is metallothionein (MT)-1G (*MT1G*) whose genes response to antioxidant response element (ARE). Nrf2 can bind with ARE to induce cytoprotective responses to oxidative stress. Therefore, the results suggest the possibility of Nrf2-mediated upregulation of MT, which may play a role in protective stress responses against OUFBW stimulation [21]. On the other hand, HES1 has been reported to cooperate with retinoblastoma protein to activate transcription factor RUNX2, which is required for osteoblast differentiation and bone formation [22]. It may be possible that upregulated HES1 is involved in osteoblast differentiation in periodontal ligaments.

10.3.2 Wound Healing Effects via Modification of Inflammation

In this section, the effects and mechanisms of wound healing of OUFBW are described. After inflicting a wound on the dorsal of mice (male, 8 weeks old; C57BL/6J), OUFBW or OUFBW inactive was placed on the wound area every day. Tissue morphometric analysis and qPCR using granulation tissue were performed [23].

The in vivo mouse model of wound healing revealed that wound healing was significantly accelerated in the OUFBW group upon morphometric analysis at 7 days after wounding (Fig. 10.7).

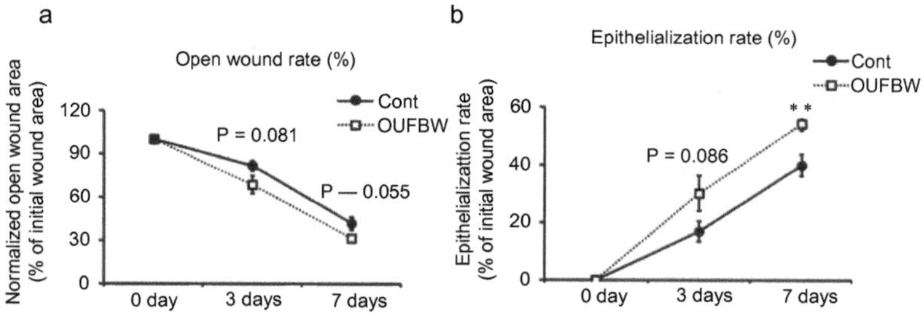

Figure 10.7 Comparison of wound healing between control and ozone ultrafine bubble water (OUFBW)-treated mice. Tissue morphometric analysis was performed in the control group and OUFBW group after 0, 3, and 7 days after administration. (a) Open wound rate (%), (b) epithelialization rate (%) (n = 7). **: $P < 0.01$.

Furthermore, mRNA expression of the inflammatory markers, tumor necrosis factor-α (TNF-α), and interleukin 6 (IL-6) mRNA, was significantly downregulated, while fibroblast growth factor-2 was upregulated in the granulation tissue of OUFBW mice after 7 days. Histological analysis revealed no sign of infection, acute inflammation, and abnormal healing in OUFBW mice after 7 days. However, TNF-α and collagen type IV mRNA expression was significantly upregulated after 3 h of OUFBW administration in the early phase of wound healing.

Ozone did not inhibit inflammation but directly induced inflammation in the early stage (on day 3) of wound healing. Three hours after OUFBW administration on day 3 of wound infliction, indicating phase I inflammation stage, TNF-α mRNA (*Tnfα*) was significantly upregulated. In addition, IL-6 mRNA (*Il6*) was slightly but not significantly upregulated in OUFBW mice. However, on day 7, indicating phase II intermediate stage, *Tnfα* and *Il6* were significantly downregulated in OUFBW-treated mice compared to those in control mice. These results indicate that OUFBW may promote inflammation during phase I of the healing process, and consequently accelerate wound healing.

Ozone generates ROS, including hydroxyl radicals, superoxide ($O_2^{\cdot-}$), and hydrogen peroxide (H_2O_2), and inflammatory mediators such as TNF-α in inflammatory cells, thus playing an integral role in modulating several physiological functions [24]. These properties are beneficial in wound healing because infection inhibits this process. Valacchi et al. reported that healing of mouse skin lesions was significantly accelerated, particularly during phase I of wound healing, i.e., epithelization, upon topical application of ozonated sesame oil [25]. Furthermore, they observed increased angiogenesis and enhanced vascular endothelial growth factor (VEGF) and cyclin D1 expression.

TNF-α is one of the inflammatory cytokines and plays an important role in numerous physiological processes. TNF-α upregulation in the early stage of wound healing might induce tissue inflammation, which is important during wound healing [26]. Combined with the high-potential sterilization effect of ozone, OUFBW plays a very effective and important role in, as it were, aseptic inflammation. Araneda et al. reported that ozone-activated immune cells stimulated the release of cytokines, including TNF-α, IL-6, and VEGF, and that VEGF upregulation persisting after ozone exposure may contribute to brain repair and consecutive functional adaptations [27].

Collagen type IV (collagen IV) is a unique component of the basement membrane (BM) in all animal phyla [28]. The BM zone has important structural and functional roles on blood vessels, constituting an extracellular microenvironmental sensor for endothelial cells and pericytes, and widely distributed extracellular matrices form an interface at the basilar surface of epithelial and endothelial cells and surround the muscle, adipose, and Schwann cells [29]. Collagen IV mRNA was upregulated 3 h after OUFBW administration at the early phase of wound healing. As this type of collagen is an important component of BM, this effect of OUFBW is beneficial for wound healing.

The mRNA of two inflammatory cytokines, *Tnfα* and *Il6,* was downregulated on day 7 in OUEBW-treated mice. At this stage, the effect of ozone on the induction of inflammatory cytokines was decreased. These results suggest that OUFBW regulates the expression of inflammatory cytokines, consistent with the requirements of wound healing. Several studies have shown that

OUFBW has antibacterial effects and induces oxidative stresses [11, 30]. OUFBW might promote wound healing via tolerance induction against oxidase stresses rather than via antibacterial effects.

FGF2 mRNA expression was induced on day 7 of OUFBW administration. FGF2 secreted by activated fibroblasts promotes vascularization [31]. Bayer et al. compared two treatment methods, viz., low-level laser therapy (LLLT) and ozone for oral mucositis, using Sprague Dawley rats and reported that LLLT and ozone may accelerate wound healing, and ozone also induces FGF and platelet-derived growth factor expression, although LLLT was more effective than ozone [32]. Interestingly, TNF-α has been reported to significantly upregulate FGF2 production in human mesenchymal stem cells [33]. Our results are partly concurrent with these results.

10.4 Epilegomena

From a clinical point of view, the validity of OUFBW is clear as an adjunctive therapy of the treatment to periodontitis, which is a bacterial inflammatory disease. Furthermore, our country is a superaged society with the highest elderly ratio in the world. Since patients under perioperative or elderly persons often ingest gargles accidentally during oral hygiene management, drinkable gargles are important and useful for those subjects. An adaptative probability of intraoral mucositis by the administration of an anticancer agent or immunosuppressive agent for bone marrow transplantation is also considered.

References

1. Haffajee, A. D. and Socransky, S. S. (2001). Relationship of cigarette smoking to the subgingival microbiota. *J. Clin. Periodontol.*, **28**, pp. 377–388.
2. Petersilka, G. J., Ehmke, B., and Flemmig, T. F. (2002). Antimicrobial effects of mechanical debridement. *Periodontol. 2000*, **28**, pp. 56–71.
3. Slots, J. (2017). Periodontitis: Facts, fallacies and the future. *Periodontol. 2000*, **75**, pp. 7–23.
4. Karpiński, T. M. and Szkaradkiewicz, A. K. (2015). Chlorhexidine – pharmaco-biological activity and application. *Eur. Rev. Med. Pharmacol. Sci.*, **19**, pp. 1321–1326.

5. Kim, J. G., Yousef, A. E., and Dave, S. (1999). Application of ozone for enhancing the microbiological safety and quality of foods: A review. *J. Food Prot.*, **62**, pp. 1071–1087.
6. Restaino, L., Frampton, E. W., Hemphill, J. B., and Palnikar, P. (1995). Efficacy of ozonated water against various food-related microorganisms. *Appl. Environ. Microbiol.*, **61**, pp. 3471–3475.
7. Huth, K. C., Jakob, F. M., Saugel, B., Cappello, C., Paschos, E., Hollweck, R. et al. (2006). Effect of ozone on oral cells compared with established antimicrobials. *Eur. J. Oral Sci.*, **114**, pp. 435–440.
8. Filippi, A. (2001). The effects of ozonized water on epithelial wound healing. *Deutsche Zahnärztliche Zeitschrift* **56**, pp. 104–108 (in German).
9. Ebensberger, U., Pohl, Y., and Filippi, A. (2002). PCNA-expression of cementoblasts and fibroblasts on the root surface after extraoral rinsing for decontamination. *Dent Traumatol.*, **18**, pp. 262–266.
10. Nagayoshi, M., Fukuizumi, T., Kitamura, C., Yano, J., Terashita, M., and Nishihara, T. (2004). Efficacy of ozone on survival and permeability of oral microorganisms. *Oral Microbiol. Immunol.*, **19**, pp. 240–246.
11. Hayakumo, S., Arakawa, S., Takahashi, M., Kondo, K., Mano, Y., and Izumi, Y. (2014). Effects of ozone nano-bubble water on periodontopathic bacteria and oral cells: In vitro studies. *Sci. Technol. Adv. Mater.*, **15**, pp. 1–7.
12. Hayakumo, S., Arakawa, S., Mano, Y., and Izumi, Y. (2013). Clinical and microbiological effects of ozone nano-bubble water irrigation as an adjunct to mechanical subgingival debridement in periodontitis patients in a randomized controlled trial. *Clin. Oral Investig.*, **17**, pp. 379–388.
13. Arakawa, S., Sugisawa, M., and Leewananthawet, A. (2017). Application of ozone nanobubble water (ONBW) to peri-implantitis treatment. *Dentistry*, **7**, pp. 1–5.
14. Lang, N. P., Berglundh, T., Heitz-Mayfield, L. J., Pjetursson, B. E., Salvi, G. E., et al. (2014). Consensus statements and recommended clinical procedures regarding implant survival and complications. *Int. J. Oral MaxillofacImplants.*, **19**, pp. 150–154.
15. Gloria, I. L., Marı´a, A. S., Diana, M. C., Marı´a, V. R., and Luz, A. G. (2017). Microbiome and microbial biofilm profiles of peri-implantitis: A systematic review. *J. Periodontol.*, **88**, pp. 1066–1089.
16. Erdemci, F., Gunaydin, Y., Sencimen, M., Bassorgun, I., and Ozler, M. (2014). Histomorphometric evaluation of the effect of systemic and

topical ozone on alveolar bone healing following tooth extraction in rats. *Int. J. Oral Maxillofac Surg.*, **43**, pp. 777–783.

17. Ozdemir, H., Toker, H., Balcı, H., and Ozer, H. (2013). Effect of ozone therapy on autogenous bone graft healing in calvarial defects: A histologic and histometric study in rats. *J. Periodont. Res.*, **48**, pp. 722–726.
18. Kazancioglu, H. O., Ezirganli, S., and Aydin, M. S. (2013). Effects of laser and ozone therapies on bone healing in the calvarial defects. *J. Craniofac. Surg.*, **24**, pp. 2141–2146.
19. Leewananthawet, A., Arakawa, S., Okano, T., Daitoku Kinoshita, R., Ashida, H., Izumi, Y., et al. (2019). Ozone ultrafine bubble water induces the cellular signaling involved in oxidative stress responses in human periodontal ligament fibroblasts. *Sci. Technol. Adv. Mater.*, **20**, pp. 589–598.
20. Kletsas, D., Basdra, E. K., and Papavassiliou, A. G. (2002). Effect of protein kinase inhibitors on the stretch-elicited c-Fos and c-Jun up-regulation in human PDL osteoblast-like cells. *J. Cell Physiol.*, **190**, pp. 313–321.
21. Motohashi, H. and Yamamoto, M. (2004). Nrf2-Keap1 defines a physiologically important stress response mechanism. *Trends Mol. Med.*, **10**, pp. 549–557.
22. Suh, J. H., Lee, H. W., Lee, J. W., and Kim, J. B. (2008). Hes1 stimulates transcriptional activity of Runx2 by increasing protein stabilization during osteoblast differentiation. *Biochem. Biophys. Res. Commun.*, **367**, pp. 97–102.
23. Watanabe, K., Ohsugi, Y., Maekawa, S., Sasaki, N., Shiba, T., Katagiri, S. et al. (2019). Ozone ultrafine bubble water improves wound healing via modification of inflammation. *J. Stomatol. Soc.*, **86**, pp. 1–11.
24. Churg, A. (2003). Interactions of exogenous or evoked agents and particles: The role of reactive oxygen species. *Free Radic. Biol. Med.*, **34**, pp. 1230–1235.
25. Valacchi, G., Lim, Y., Belmonte, G., Miracco, C., Zanardi, I., Bocci, V., and Travagli, V. (2011). Ozonated sesame oil enhances cutaneous wound healing in SKH1 mice. *Wound Repair Regen.*, **19**, pp. 107–115.
26. Hao, K., Li, Y., Feng, J., Zhang, W., Zhang, Y., Ma, N., Zeng, Q., et al. (2015). Ozone promotes regeneration by regulating the inflammatory response in zebrafish. *Int. Immunopharmacol.*, **28**, pp. 369–375.
27. Araneda, S., Commin, L., Atlagich, M., Kitahama, K., Parraguez, V. H., Pequignot, J. M., et al. (2008). VEGF overexpression in the astroglial

cells of rat brainstem following ozone exposure. *Neurotoxicology*, **29**, pp. 920–927.

28. Fidler, A. L., Vanacore, R. M., Chetyrkin, S. V., Pedchenko, V. K., Bhave, G., Yin, V. P., et al. (2014). A unique covalent bond in basement membrane is a primordial innovation for tissue evolution. *Proc. Natl. Acad. Sci. USA.*, **111**, pp. 331–336.
29. Abreu-Velez, A. M. and Howard, M. S. (2012). Collagen IV in normal skin and in pathological processes. *N. Am. J. Med. Sci.*, **4**, pp. 1–8.
30. Bocci, V., Borrelli, E., Travagli, V., and Zanardi, I. (2009). The ozone paradox: Ozone is a strong oxidant as well as a medical drug. *Med. Res. Rev.*, **29**, pp. 646–682.
31. Clark, R. A., Ghosh, K., and Tonnesen, M. G. (2007). Tissue engineering for cutaneous wounds. *J. Invest. Dermatol.*, **127**, pp. 1018–1029.
32. Bayer, S., Kazancioglu, H. O., Acar, A. H., Demirtas, N., and Kandas, N. O. (2017). Comparison of laser and ozone treatments on oral mucositis in an experimental model. *Lasers Med. Sci.*, **32**, pp. 673–677.
33. Crisostomo, P. R., Wang, Y., Markel, T. A., Wang, M., Lahm, T., and Meldrum, D. R. (2008). Human mesenchymal stem cells stimulated by TNF-alpha, LPS, or hypoxia produce growth factors by an NF kappa B- but not JNK-dependent mechanism. *Am. J. Physiol. Cell Physiol.*, **294**, pp. C675–C682.

Chapter 11

Preservability of Ultrafine Bubbles

Wataru Kanematsu, Toru Tuziuti, and Kyuichi Yasui
National Institute of Advanced Industrial Science and Technology (AIST), Nagoya 463-8560, Japan
w.kanematsu@aist.go.jp

Information about the preservability of ultrafine bubbles (UFBs) dispersed in bulk pure water has been summarized. As a benchmark, temporal variations in the properties of UFBs stored under preferred conditions, in a virtually air-tight glass bottle, without any gas–liquid interface, at a constant temperature, were measured periodically with a nanoparticle tracking analyzer. The UFB number concentration slowly diminished with time and remained at 84% of its initial value of the storage during as long as over 9 months. Pronounced decrease in the concentration after freeze–thaw processing inferred that gas-filled bubbles indeed remained after such a long storage. In a realistic way of storage, the presence of air above the water surface inside the glass jar enhanced the decreasing rate of UFBs. A UFB dispersion stored in a polyethylene (PE) pouch showed a more rapid decay of the concentration compared to samples stored in a glass bottle. To have qualitative evidence about the interaction between

Ultrafine Bubbles
Edited by Koichi Terasaka, Kyuichi Yasui, Wataru Kanematsu, and Nobuhiro Aya

ISBN 978-981-4877-59-6 (Hardcover), 978-1-003-14195-2 (eBook)
www.jennystanford.com

polymer materials for container and UFBs, temporal changes in the number concentration of UFBs were measured when the materials were immersed in UFB water stored under the preferred condition. Three different polymer materials, including PE, exerted differing influences on UFBs. UFBs generated by two different generation principles exhibited difference in decreasing trends of number concentration with elapsed time.

11.1 Introduction

Many experimental results supporting the extraordinarily long lives of submicron gas-filled bubbles in bulk water, which are specified as ultrafine bubbles (UFBs) in ISO international standards [1], have been obtained through a variety of measurement techniques. For instance, UFBs were distinguished from solid particles by using a resonant mass measurement technique based on density differences between them [2]. Observations of UFBs were made with both field emission scanning electron microscopy and transmission electron microscopy [3, 4]. To explain the unusual longevity of UFBs, various models have been proposed and most of them were surveyed elsewhere [5, 6]. At this point, the dynamic equilibrium model for bulk UFBs [7] seems to hold a predominant position compared with other models because the observation of UFBs with specially designed TEM [8] lends support to the hypothesis that hydrophobic materials are adsorbed on the surface of bubbles.

To utilize the prolonged stability of UFBs, which is much longer than that predicted by classic theory [9], many attempts have been made in various areas of technology such as wastewater treatment [10], defouling of membranes [11], machining fluid for metal cutting and grinding [12], vegetable seed germination [13], and irrigation water in tomato-growing operations [14, 15]. To ensure further development in these fields, the understanding of the lifetime of UFBs and its influencing factors is necessary. In this chapter, for prospective users of UFBs, information about the preservability of UFBs has been summarized as follows. First as a benchmark, temporal variations in the properties of UFBs, such as number concentration and mean diameter, were monitored under preferred storage conditions; a storage container is virtually air-tight, and temperatures are kept constant at 25°C in a vibration-free environment to eliminate the factors that presumably cause the disappearance of UFBs. Then, the influence of interaction between UFBs and surrounding air in a container on number concentration was examined along with the effect of aeration with gas-washing bottle. Interaction with polymer materials commonly used for storage of water containing UFBs was also examined. Finally, the difference in temporal changes in the concentration of UFBs between two different generation principles was investigated.

11.2 Generation of Ultrafine Bubbles

To suppress the influence of contamination in raw water, UFBs were generated in pure water (pH = 6.81±0.09), prepared by a Millipore Elix Essential 5 system (Merck, USA). The bubbles were produced using two types of commercially available bubble generators based on different principles. One generates UFBs by hydrodynamic cavitation followed by pulverization by shear force in vortex flow [16]. This type of generator (Type A) was mainly used in the following. The number concentration of bubbles increased with the number of times the water volume circulated through the generator, which is called pass as shown in Fig. 11.1. The other (Type B) generates UFBs by pressurized dissolution of air followed by steep pressure reduction and consecutive bubble releasing through a nozzle [17]. The number

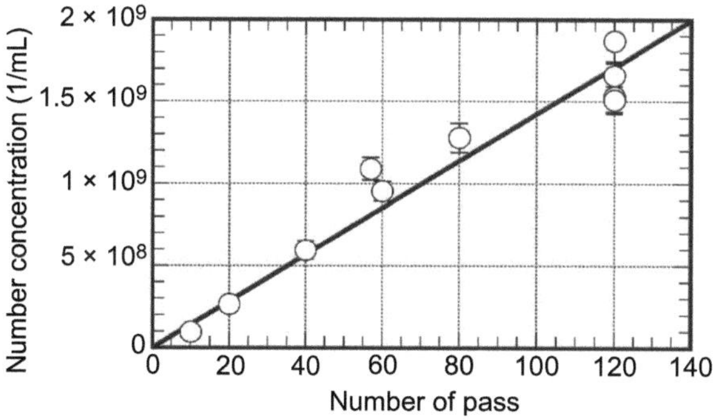

Figure 11.1 Increase in number concentration with increase in number of pass during UFB generation. Error bars indicate the standard deviation in 3 to 5 consecutive measurements as is the case with the following figures.

concentration of bubbles by Type B generator also increased with increase in the number of pass. In both types of generators, self-primed air was filtered with a hollow fiber membrane to prevent any mixing of atmospheric particulate matter. In this chapter, UFBs were generated by 120 passes from the pure water. The temperatures of samples by Type A and B generators reached around 50 and 60°C, respectively, after the generation.

11.3 Characterization of UFBs

The size distributions and number concentrations of bubbles were characterized with a Nanosight NS300 instrument (Malvern Panalytical, UK), which operates using the nanoparticle tracking analysis (NTA) principle. The instrument was equipped with a blue laser light source (405 nm, 65 mW), high sensitivity CMOS camera, 20× magnification microscope, and integrated syringe pump. The data collection and analysis were performed with software NTA 3.2. In Nanosight, the number concentration is calculated from the number of light-scattering centers located in the focal plane of the microscope, adjusted to a per mL basis. The bubble size is evaluated by recording the two-dimensional Brownian motion of the individual light-scattering centers in the microscope's field of view and calculating the particles' hydrodynamic diameters based on the Einstein–Stokes equation [18]. Sample volume in the instrument is extremely small (approximately 80 pico liters; 100 by 80 μm optical field of view and 10 μm depth of illuminating beam), and a multiplication factor of 1.25×10^7 is required to convert the analyzed value to a conventional unit of number concentration of bubbles (particles per mL). Small changes in the number of bubbles recorded per frame are greatly magnified in the analysis. Unless otherwise specified, the measurement and analysis were conducted under laminar flow using an integrated syringe pump under the control of the NTA software to reduce the influence of such magnification on the concentration value. For avoiding the contamination during infusion, we adopted a syringe that is not made with natural rubber latex and silicone oil free. To extract the Brownian motion of individual bubbles, the translational motion of the bulk liquid sample is subtracted from the overall motion by the software. Samples were infused to the measurement cell at a rate of approximately 20 μL/min during the measurement. Some samples were infused manually by using the same type of syringe mentioned earlier. The length of video acquisition for each measurement was 60 s under both infusing manners, propelling UFB water by syringe with pump or manually.

11.4 Storage of UFBs Dispersed in Bulk Water

Each fresh UFB water was cooled slowly to room temperature. The concentration of dissolved oxygen reduced from supersaturation level to a saturated one during the cooling. Then the water was poured into two types of glass container and a polymer pouch. One of the glass containers is a glass (DURAN®, Schott AG, Germany) bottle with a screw-on cap having a gasket made of polybutylene terephthalate (PBT) attached to the inside, in which permeation of gas and water vapor is effectively impeded. The nominal volume of the bottle is 25 mL. The other is a screw-on cap glass jar with snap-in inner lid made of PE and its nominal volume is 100 mL. The gas barrier property of PE is not so much as that of PBT. The polymer pouch with a screw-on cap was made of PE film laminated with an outer film of nylon. Unless otherwise specified, the containers include no gas–liquid boundary inside and were stored at a constant temperature of 25°C, controlled with a Peltier device. These storing conditions enable to reduce the factors of the equilibrium of UFB water and to provide test environment suitable for the influence of long-term storage on the properties of UFBs eventually. Hereafter we call this air-tight and constant temperature (ATCT) storage, as contrasted with that open to the atmosphere. The containers were opened immediately before Nanosight measurement. In the following sections, the elapsed time in chronological changes in UFB properties originates from the beginning of storage, not the generation of UFBs.

11.5 Chronological Changes in UFB Properties during ATCT Storage

11.5.1 Number Concentration

The number concentration of UFBs at the bottling varied from batch to batch, even when the number of pass was kept constant; the concentration ranged from 1.15×10^9 to 1.58×10^9 particles per mL. To compare the temporal variation between batches, the number concentrations were normalized by the initial value of each batch. The normalized bubble concentration of seven batches varied as shown in Fig. 11.2. Results of linear regression analysis by the least-square approach on a log–log scale and its 95% confidential intervals are also shown in the figure. Compared to existing reports [19–21], the concentration diminished more slowly with time. Even after storage of as long as 9 months, the bubble concentration remained at 84% of its initial value on average. The small but steady decrease in number concentration with time, which would be unlikely in solid particles, infers that air-filled bubbles are indeed the most of nanoentities observed by Nanosight. Considering the fact that no visible macroscopic bubbles appeared in the bottle as a consequence of long-term storage, the coalescence of the UFBs, which should lead to the formation of microbubbles, is unlikely. Other mechanisms such as dissolution of air inside the bubbles by internal pressure, and hydrophobic interaction between UFBs and PBT gasket, could bring about the slow decrease in the number concentration.

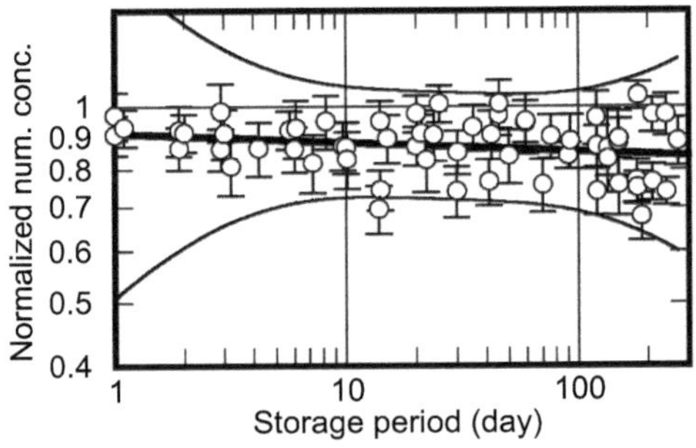

Figure 11.2 Temporal variation in normalized number concentration of ultrafine bubbles stored in glass bottle under air-tight conditions at 25°C. Reprinted with permission from Kanematsu [22]. Copyright (2020), Elsevier.

11.5.2 Verification of Existence of Gas-Filled Bubbles

The existence of UFBs can be verified by slow freezing of the aqueous sample followed by thawing at room temperature (freeze–thaw process) [19, 23]. Although the freeze–thaw process is rather time consuming compared to other processes such as mixing with degassed water [16] or ultrasonic irradiation [24], drastic reductions in UFBs through this process was reported by some researchers [19, 23, 25]. Approximately 20 mL of UFB sample in a glass bottle of nominal volume 25 mL with a screw-on cap was frozen slowly and kept at around −12°C for 10 h, then followed to thaw at room temperature before analysis with Nanosight. The process was applied to samples stored for 1, 10, 120, and 209 days. The decreasing rate of bubble number concentration reached over 90%, and such a high rate was kept after the elapsed time of as long as 209 days [22]. The number concentration was converted to a volume concentration using the assumption that all detected bubbles or particles were spherical in shape. From the summation in the volume of spheres over diameter, the decrease in volume was estimated to be around 60%. It seems to be unlikely that such a significant reduction in volume was caused by agglomeration of solid particles alone.

Considering the results described earlier, UFBs appear to have survived after storage for more than 9 months under the ATCT condition. Most of the nanoparticles observed with Nanosight were inferred to be air-filled nanobubbles.

11.5.3 Mean Diameter

Mean diameters were normalized by the initial value for each batch as with the case of number concentration. Temporal variations in the normalized mean diameter of seven batches are summarized in Fig. 11.3. The normalized values slightly increased with time up to around 70 days and then marginally accelerated with wide variation from sample to sample. In the comparison of bubble size distributions between the bottling day and specific days elapsed, selective loss of bubbles smaller than the initial modal diameter was observed. Meanwhile, no clear trend was observed in the size distributions of bubbles larger than the mode (not shown here). Similar preferential decrease in smaller bubbles was reported [16], and it was inferred that the gas phase inside a smaller bubble will more readily dissolve into the water due to the increased Laplace pressure. Considering the fact that the number concentration decreased monotonically with time, as shown in Fig. 11.2, the observed marginal acceleration after 70 days suggests that residue, such as solid particles and liquid droplets present after complete dissolution of gas inside the bubbles, may have agglomerated and formed particles that were detected by Nanosight.

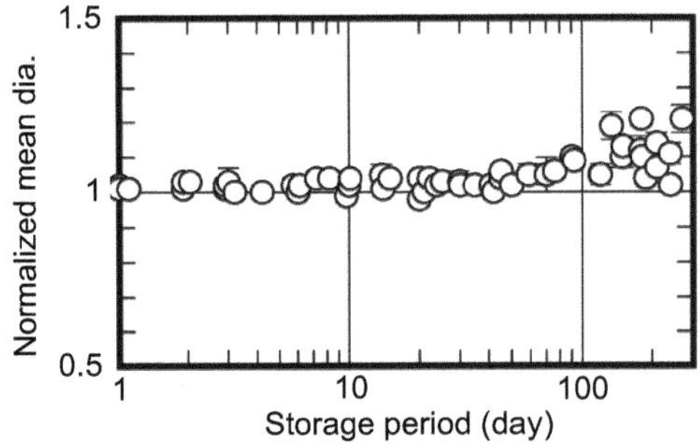

Figure 11.3 Temporal variations in mean diameters normalized by the initial value for each generation batch. Reprinted with permission from Kanematsu [22]. Copyright (2020), Elsevier.

11.5.4 Zeta Potential

For some samples, the zeta potential of UFBs was measured using an electrophoresis-type instrument, the Zetasizer Nano ZS (Malvern Panalytical, UK). Zeta potential measurements were performed nearly simultaneously with the Nanosight measurements. In contrast to number concentration and mean diameter, zeta potential did not show pronounced temporal variation and ranged from −35 to −17 mV, even in a sample stored for more than 6 months, as shown in Fig. 11.4. The trend in the range of zeta potential was of a similar level as those reported elsewhere [19, 20, 26], where UFBs were found to be negatively charged. Meanwhile, zeta potential decreased to an unmeasurable level after the freeze–thaw process of samples, coinciding with the drastic decrease in number concentration of UFBs. These results inferred that negative charge of bubbles contributes to their stabilization.

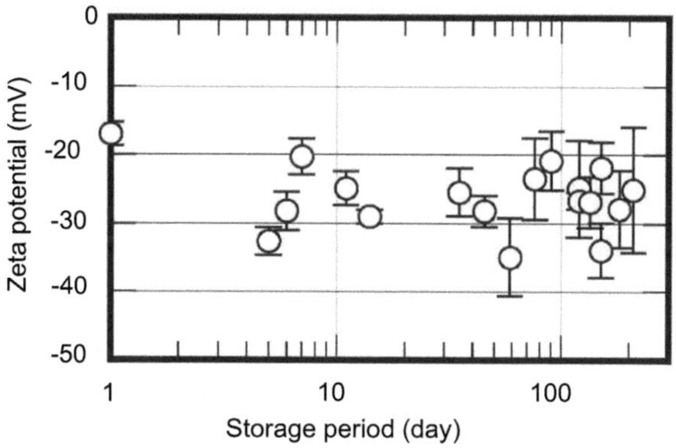

Figure 11.4 Temporal variations in zeta potential of UFBs stored in glass bottle under air-tight conditions at 25°C.

11.6 Influence of Storing Conditions on Chronological Changes in UFB Properties

11.6.1 Influence of Air inside Container

Figure 11.5 exhibits that the chronological change in number concentrations of UFBs stored in a glass jar were compared between two different storing conditions; UFB water was stored with and without air above the water surface therein. The jars were stored at a constant temperature of 25°C, controlled with a Peltier device as is the case with the ATCT storage. For storing without air, the concentration decreased with time slowly comparable to the decreasing rate under ATCT condition in Fig. 11.2. After the storage of 60 days, the concentration remained at more than 80% of its initial value. Meanwhile, for storing with air above the water surface, the decrease became marked after a lapse of 10 days and the concentration reached as low as 30% of its initial value eventually after 300 days. It is quite likely that interaction between UFBs and air at the water surface exerts a significant influence on diminishing of UFBs.

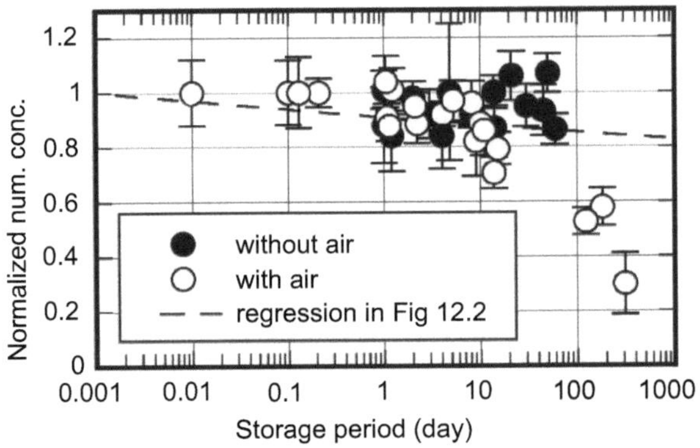

Figure 11.5 Temporal variations in number concentration of UFBs. UFB water was stored in glass jar with and without air therein.

11.6.2 Influence of Aeration

To bring air into contact with UFBs more actively, aeration with gas-washing bottle was conducted. The UFB water was stirred with bubbles of visible size discharged from the nozzles of Muenck-type gas-washing bottle. Most UFBs stored in the bottle were desired to contact with these air bubbles. As an index of contact frequency between UFBs and air, air flow per unit volume of UFB water (L/mL) was introduced, as shown in Fig. 11.6. With increase in contact frequency, normalized number concentration decreased toward the value range of 0.6 to 0.7. The effect of diminishing UFBs by contact with air flow exists indeed. However, the amount of decrease in number concentration is plateaued. Air through the gas-washing bottle was not filtered because an air pump does not provide enough discharge pressure to install a filter at pump outlet. It seems likely that the increase in the number of dusts in air counted as particles by Nanosight balances the decrease in UFBs by contact with air flow. In principle, it is impossible to distinguish solid particles from air-filled bubbles by an instrument based on the particle tracking analysis method such as Nanosight. For further quantitative examination, clean air is desired in aeration.

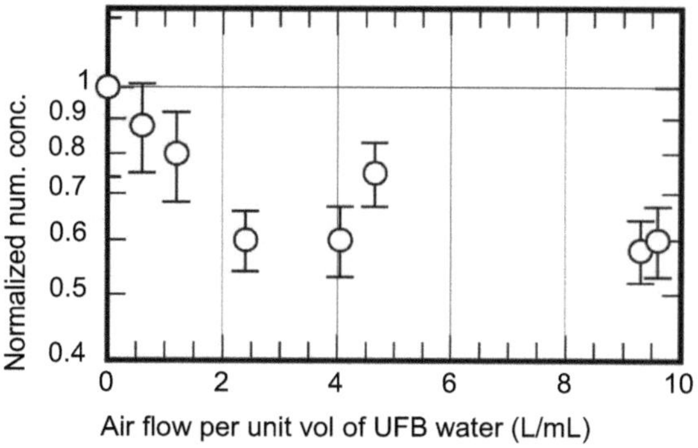

Figure 11.6 Diminishing UFBs by contact with air flow with Muenck-type gas-washing bottle.

11.7 Influence of Container Materials on Chronological Changes in UFB Properties

11.7.1 Preserving Property of Polymer Pouch in Storage of UFB Water

As an example of typical storage for UFB water, a polymer pouch with a screw-on cap was adopted. The pouch was made of PE film laminated with an outer film of nylon. Storage conditions were the same as the glass bottle; however, the gas permeability of PE is much higher than that of PBT, presumably leading to much higher gas diffusion through the polymer pouch than the glass bottle. The UFB water was stored in the PE pouch for up to 24 days under conditions similar to ATCT storage, and the temporal change in the properties of the water was examined.

As with the case of UFB water stored in the glass bottle, chronological changes in normalized number concentration for three batches of the water stored in the pouch are plotted in Fig. 11.7. After 5 days, the concentration sharply declined and reached as little as 0.1 after 20 days. The cause of such sharp decline will be discussed in the following.

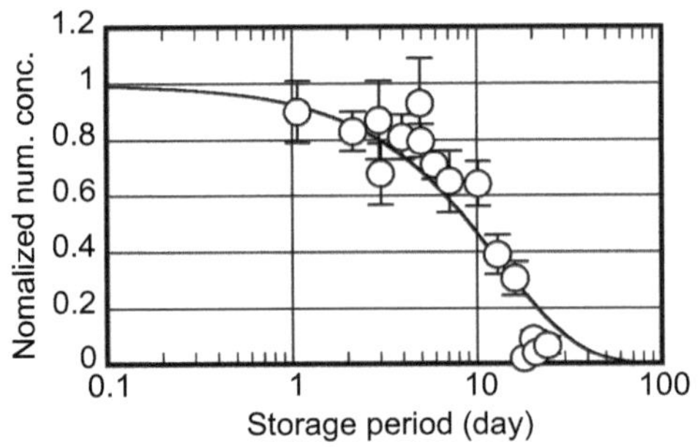

Figure 11.7 Temporal variation in normalized number concentration of UFB water stored in a PE pouch. Reprinted with permission from Kanematsu [22]. Copyright (2020), Elsevier.

11.7.2 Interaction between Container Materials and UFBs

To have qualitative information about the influence of container materials on the lifetime of UFBs, temporal changes in the number concentration of UFBs were measured when the materials were immersed in UFB water stored in the same glass bottle as that used for ATCT storage. Three types of polymer samples, a PE tube (2 mm ID × 4 mm OD × 60 mm in length), a nylon ball (6.4 mm in diameter), and a polyethylene terephthalate (PET) strip, were examined to evaluate the influence of interaction between UFB surfaces and polymer materials on the properties of UFBs. The area of contact per unit volume of UFB water was adjusted to match that of the PE pouch (2.88 cm^2/mL) described in Section 11.7.1 for the comparison afterward.

During the ATCT storage, the changes should be due to interactions between polymer materials and surrounding UFBs. Even under the ATCT condition, the number concentration slightly decreased with time, as described in Section 11.5.1. Therefore, the deviation from the regression value for UFB water stored under ATCT condition, expressed as the straight line on a log–log scale in Fig. 11.2, is considered to indicate the net influence of interaction. Hereafter we call the deviation the net number concentration (NNC). Figure 11.8 exhibits the temporal variations in the NNC up to

28 days elapsed. The trend in each material is marked by black lines. The concentrations decreased with time, particularly over the first 5 days, followed by slower decreases out to 30 days of storage. The decreasing UFB concentrations exposed to PET and PE seem to level off after about 10 days. After a storage of 20 days, concentrations of bubbles exposed to the three materials were in the following order: PET, PE, nylon. Based on the dynamic equilibrium model [7], possible causes of such decrease are presumed as follows: Bubbles were adsorbed on the surfaces of the polymers, and the number of bubbles in the bulk liquid was thereby reduced. The adsorption suggests the presence of hydrophobic interaction between substances covering UFB surface and polymer materials.

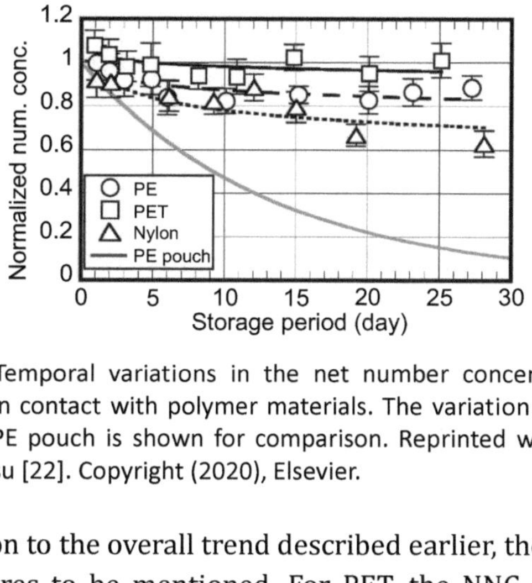

Figure 11.8 Temporal variations in the net number concentration (NNC) of UFB water in contact with polymer materials. The variation for UFB water stored in the PE pouch is shown for comparison. Reprinted with permission from Kanematsu [22]. Copyright (2020), Elsevier.

In addition to the overall trend described earlier, there are some unique features to be mentioned. For PET, the NNC ranged from 0.934 to 1.078 and its average was 0.993, demonstrating that the decrease in the NNC for PET was relatively small. It suggests the potential of PET as a container, which can maintain UFB water with small losses at the same level as those in glass containers. Whereas nylon exhibited pronounced decreases in the NNC, and the reduction reached around 40% at 28 days, implying less efficiency in preserving UFBs than PET.

The regression curve in Fig. 11.7 was compared with the decrease in NNC in Fig. 11.8. The decrease in UFB concentration after storage in the PE pouch (the gray thin line in the figure) was much more than that of NNC by interaction between bubble surfaces and a PE tube. Generally, the gas permeability of PE is much greater than that of the PBT used for the gasket in the screw-on glass bottle cap. Also, it is widely accepted that the gas barrier performance of nylon, which was used as the outer film of the pouch, is potentially deteriorated by exposure to water vapor [27]. In these circumstances, gas in bulk water and water vapor can diffuse to the inner wall of the PE pouch and the water vapor passing through the PE film can deteriorate the outer nylon film. Eventually, gas molecules in UFBs adsorbed onto the inner pouch wall by hydrophobic interaction have the potential to pass through to the atmosphere. This could lead to a sharp decrease in bulk UFBs, consistent with the regression curve in Fig. 11.7, which shows an exponential decrease in bubble concentration with time. Consequently, it is suggested that the diffusion of gas to the atmosphere is predominant in the decrease in UFBs stored in this PE pouch.

11.8 Difference in Temporal Change of Number Concentration of UFBs between Different Generation Principles

The chronological change in number concentrations of UFBs generated by Type B generator is shown in Fig. 11.9. The regression curve for UFBs by Type A generator was indicated by the dashed line in the figure. After the elapse of 5 days, decrease in number concentration became marked and the concentration reached as low as 60 to 70% of its initial value eventually after 30 days elapsed. These demonstrated that temporal change of number concentration differs from UFBs generated by different generation principle. Given that the dynamic equilibrium model is also the case with UFBs by Type B generator, the difference in diminishing rate implies that the kinds of substance on bubble surface or adsorption force are different from those of UFBs by Type A generator.

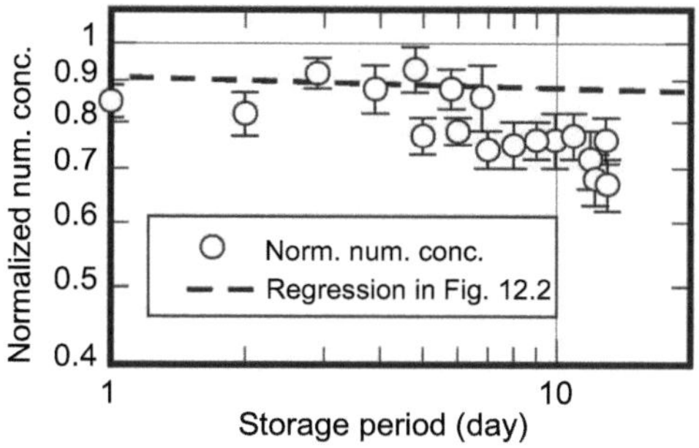

Figure 11.9 Temporal change in number concentration of UFBs generated by Type B generator.

11.9 Summary

Information about the preservability of ultrafine bubbles (UFBs) dispersed in bulk pure water has been summarized. As a benchmark, the UFB water was stored under conditions enable to reduce the factors of the equilibrium of UFB water and to provide test environment suitable for the influence of long-term storage on the properties of UFBs; no gas–liquid boundary inside, vibration-free, and a constant temperature of 25°C controlled with a Peltier device. The number concentration of UFBs slowly diminished with time, and the concentration remained at 84% of initial value after the storage of as long as over 9 months. Pronounced decrease in the number concentration after freeze–thaw processing inferred that gas-filled bubbles indeed remained even after such a long storage. In a realistic way of storage, the presence of air above the water surface inside the glass jar enhanced the decreasing rate of UFBs, and its supporting evidence was provided by the aeration of UFB water. A UFB dispersion stored in a PE pouch showed a more rapid decay of the number concentration compared to storage in a glass bottle, inferring the presence of interaction between bubble surface and container materials in conjunction with diffusion of gas molecules through the pouch wall. To have qualitative evidence for such interaction, temporal changes in the number concentration of UFBs were measured when the materials were immersed in UFB water stored under the preferred condition. Three different polymer materials, including PE, exerted differing influences on the concentrations. UFBs generated by two different generation principles exhibited differing decreasing trends of number concentration with elapsed time, implying that the kinds of substance on bubble surface or adsorption force are different from each other.

Acknowledgments

The authors would like to thank Tomoko Tanaka for meticulous measurements of UFBs.

References

1. ISO 20480-1:2017. Fine bubble technology: General principles for usage and measurement of fine bubbles — Part 1: Terminology.
2. Kobayashi, H., Maeda, S., Kashiwa, M., and Fujita, T. (2014). Measurements of ultrafine bubbles using different types of particle size measuring instruments. In: *Proc. SPIE 9232*, p. 92320U. https://doi.org/10.1117/12.2064638.
3. Ohgaki, K., Khanh, N. Q., Joden, Y., Tsuji, A., and Nakagawa, T. (2010). Physicochemical approach to nanobubble solutions. *Chem. Eng. Sci.*, **65**, pp. 1296–1300. https://doi.org/10.1016/j.ces.2009.10.003.
4. Uchida, T., Liu, S., Enari, M., Oshita, S., Yamazaki, K., and Gohara, K. (2016). Effect of NaCl on the lifetime of micro- and nanobubbles. *Nanomaterials*, **6**, 31. https://doi.org/10.3390/nano6020031.
5. Alheshibri, M., Qian, J., Jehannin, M., and Craig, V. S. J. (2016). A history of nanobubbles. *Langmuir*, **32**, pp. 11086–11100. https://doi.org/10.1021/acs.langmuir.6b02489.
6. Yasui, K., Tuziuti, T., and Kanematsu, W. (2018). Mysteries of bulk nanobubbles (ultrafine bubbles); stability and radical formation. *Ultrasonics Sonochemistry*, **48**, pp. 259–266. https://doi.org/10.1016/j.ultsonch.2018.05.038.
7. Yasui, K., Tuziuti, T., Kanematsu, W., and Kato, K. (2016). Dynamic equilibrium model for a bulk nanobubble and a microbubble partly covered with hydrophobic material. *Langmuir*, **43**, pp. 11101–11110. https://doi.org/10.1021/acs.langmuir.5b04703.
8. Sugano, K., Miyoshi, Y., and Inazato, S. (2017). Study of ultrafine bubble stabilization by organic material adhesion. *Jpn. J. Multiphase Flow*, **31**, pp. 299–306.

9. Epstein, P. S. and Plesset, M. S. (1950). On the stability of gas bubbles in liquid–gas solutions. *J. Chem. Phys.*, **18**, pp. 1505–1509.
10. Agarwal, A., Ng, W. J., and Liu, Y. (2011). Principle and applications of microbubble and nanobubble technology for water treatment. *Chemosphere*, **84**, pp. 1175–1180. https:// doi.org/10.1016/j.chemosphere.2011.05.054.
11. Ghadimkhani, A., Zhang, W., and Marhaba, T. (2016). Ceramic membrane defouling (cleaning) by air nano bubbles. *Chemosphere*, **146**, pp. 379–384. http://dx.doi.org/10.1016/j.chemosphere.2015.12.023
12. Kobayashi, H., Kamijyo, Y., Hirano, M., and Araki, K. (2020). Ultrafine bubble generation technique for machining fluid and improving efficiency of grinding (in Japanese), *Proc. Spring Meeting of JSPE*, pp. 717–718.
13. Liu, S., Oshita, S., Kawabata, S., Makino, Y., and Yoshimoto, T. (2016). Identification of ROS produced by nanobubbles and their positive and negative effects on vegetable seed germination. *Langmuir*, **32**, pp. 11295–11302. https://doi.org/10.1021/acs.langmuir.6b01621.
14. Uchida, T., Nishikawa, H., Sakurai, N., Asano, M., and Noda, N. (2018). Ultra-fine bubble distributions in a plant factory observed by transmission electron microscope with a freeze-fracture replica technique. *Nanomaterials*, **8**, 152. http://doi.org/10.3390/nano8030152.
15. Mochizuki, Y., Zhao, T., Kanematsu, W., Kawasaki, T., Saito, T., Ohyama, A., Nakano, A., and Higashide, T. (2019). Application of a growth model to validate the effects of an ultrafine-bubble nutrient solution on dry matter production and elongation of tomato seedlings. *Horticulture J.*, **88**, pp. 380–386. https://doi.org/10.2503/hortj.UTD-055.
16. Tuziuti, T., Yasui, K., and Kanematsu, W. (2018). Influence of addition of degassed water on bulk nanobubbles. *Ultrasonics Sonochemistry*, **43**, pp. 272–274. https://doi.org/10.1016/j.ultsonch.2018.01.015.
17. Terasaka, K., Himuro, S., Ando, K., and Hata, T. (2016). *Introduction to Fine Bubble Science and Technology*, edited by the Union of Fine Bubble Scientists and Engineers (Nikkan Kogyo Shimbun, Ltd, Tokyo) (in Japanese).
18. Patois, E., Capelle, M. A. H., Palais, C., Gurny, R., and Arvinte, T. (2012). Evaluation of nanoparticle tracking analysis (NTA) in the characterization of therapeutic antibodies and seasonal influenza vaccines: Pros and cons. *J. Drug Del. Sci. Tech.*, **22**, pp. 427–433.

19. Nirmalkar, N., Pacek, A. W., and Barigou, M. (2018). On the existence and stability of bulk nanobubbles. *Langmuir*, **34**, pp. 10964–10973. https://doi.org/10.1021/acs.langmuir.8b01163.

20. Zhu, J., An, H., Alheshibri, M., Liu, L., Terpstra, P. M. J., Liu, G., and Craig, V. S. J. (2016). Cleaning with bulk nanobubbles. *Langmuir*, **32**, pp. 11203–11211. https://doi.org/10.1021/acs.langmuir.6b01004.

21. Ke, S., Xiao, W., Quan, N., Dong, Y., Zhang, L., and Hu, J. (2019). Formation and stability of bulk nanobubbles in different solutions. *Langmuir*, **35**, pp. 5250–5256. https://doi.org/10.1021/acs.langmuir.9b00144.

22. Kanematsu, W., Tuziuti, T., and Yasui, K. (2020). The influence of storage conditions and container materials on the long-term stability of bulk nanobubbles: Consideration from a perspective of interactions between bubbles and surroundings. *Chem. Eng. Sci.*, **55**, pp. 308–312.

23. Nirmalkar, N., Pacek, A. W., and Barigou, M. (2019). Bulk nanobubbles from acoustically cavitated aqueous organic solvent mixtures. *Langmuir*, **35**, pp. 2188–2195. https://doi.org/10.1021/acs.langmuir.8b03113.

24. Yasuda, K., Matsushima, H., and Asakura, Y. (2019). Generation and reduction of bulk nanobubbles by ultrasonic irradiation. *Chem. Eng. Sci.*, **195**, pp. 455–461. https://doi.org/10.1016/j.ces.2018.09.044.

25. Nirmalkar, N., Pacek, A. W., and Barigou, M. (2018). Interpreting the interfacial and colloidal stability of bulk nanobubbles. *Soft Matter*, **14**, pp. 9643–9656. https://doi.org/10.1039/c8sm01949e.

26. Jin, J., Feng, Z., Yang, F., and Gu, N. (2019). Bulk nanobubbles fabricated by repeated compression of microbubbles. *Langmuir*, **35**, pp. 4238–4245. https://doi.org/10.1021/acs.langmuir.8b04314.

27. Katsura, T. (2008). Gas-barrier polymers for packaging materials (in Japanese). *Kobunshi*, **57**, pp. 974–977.

Index